ESSENTIAL
AC COBRA

ESSENTIAL
AC COBRA

THE CARS AND THEIR STORY 1962-67
RINSEY MILLS

Published 1997 by Bay View Books Ltd
The Red House, 25-26 Bridgeland Street,
Bideford, Devon EX39 2PZ, UK

© Copyright 1997 Rinsey Mills

All rights reserved. No part of this publication
may be reproduced or transmitted in any form
or by any means, electronic or mechanical,
including photocopying, recording or in
any information storage or retrieval
system, without the prior written
permission of the publisher.

Edited by Mark Hughes
Typesetting and design by Chris Fayers & Sarah Ward

ISBN 1 870979 85 0
Printed in Hong Kong

CONTENTS

BEGINNINGS
6

LEAF-SPRUNG COBRAS
13

RACING IN THE USA
26

RACING IN EUROPE
32

THE WORLD TITLE CAMPAIGN
44

COIL-SPRUNG COBRAS
62

AFTER THE COBRA
75

APPENDIX
79

BEGINNINGS

It was an inhospitable and somewhat gloomy place, besides which the January weather was an unpleasant contrast to his homeland, but the tall Texan could not resist grinning to himself as he threw the car into another corner. A four-wheel drift became a brief scrabble for grip from the rear tyres as he powered up the straight past the bystanders gathered by the pits.

The place was Silverstone, the man Carroll Shelby. And the car? That was the first Cobra.

The onlookers were for the most part personnel from AC and wellwishers, but way out on the far side of the track two teenagers, attracted by the sounds, had hopped across a couple of fences to get a better view of the car that was circulating so quickly. The older one knew it was an AC because a friend's father had just bought one of the new Ford-engined Aces, but this car was different…

It was that sound: an uneven, hobbling beat which, as

Although powered by the venerable AC 2-litre engine, the first Aces – this is chassis number AE 72 which left the factory in May 1955 – were quite quick cars for their day. The same early nose style is seen on brochure cover.

6

BEGINNINGS

Fitting the short-stroke Ford Zephyr motor in the RS 2.6 Ace – this is chassis number RS 5020 which left the factory in September 1962 – allowed the nose to be re-styled into this lower, more aggressive form that was carried through on the Cobra.

The late David Purley (in sheepskin jacket) ran a 289 Cobra in his early days in motor racing. He and the author were also together at Silverstone early in 1962 to witness the first Cobra test session…

the revs rose, gathered itself into an enveloping roar that peaked in an urgent, hollow snarl. Furtively they approached the track to get closer and let that wonderful noise engulf them. Then, just as abruptly as it had started, the sound ceased as the testing was completed.

Those who were there that day witnessed the genesis of one of the most charismatic cars in history. Shelby returned swiftly to the US with news of how the prototype Cobra had gone, while the car returned to the AC factory at Thames Ditton, Surrey, for some more work before it could be sent across the Atlantic. And the two boys?

Neither of them ever forgot that haunting sound. In time the older boy, David Purley, became a racing driver, cutting his teeth on a 289 Cobra before graduating to Grand Prix cars. The younger one was me.

The gestation period of the Cobra had begun a good while before Shelby ever thought of putting American V8 power into an English sports car when AC Cars Ltd, with roots reaching back to before the First World War, had acquired the rights to build a sports racing car designed by John Tojeiro. This the company put into production in 1953 as the AC Ace after tidying up the body design and fitting the venerable 2-litre six-cylinder overhead camshaft engine that had powered its products since the early 1920s. The Ace's light weight and excellent roadholding more than compensated for its relative lack of power – in fact it was no slouch for its day with a 0-60mph time of around 12sec and a top speed of over 100mph.

The chassis consisted of a pair of 3in diameter steel tubes with vertical, triangulated box sections at either end and a single tubular cross-member. These fabricated box sections were flat on top so that the transverse springs for the independent suspension could be mounted across

7

ESSENTIAL AC COBRA

Two key men in the Cobra story. Carroll Shelby (left) was an outstandingly successful sports car driver who also mastered single-seaters: with this Ferrari he took the class record at Mount Washington hillclimb. Derek Hurlock (below), never one to put himself in the public eye, poses a little awkwardly for a snapshot at the front of the Thames Ditton showroom; a pair of export cars, an Ace and an early Cobra, are in the background.

them; in addition the rear one was constructed in such a way that it could double as a differential carrier. The bottom wishbones were articulated from mountings welded to the main chassis tubes, at first carrying bronze bushes but thereafter rubber ones. To provide stiffening for the aluminium body that was to be mounted thereon, the chassis also had a 1½in diameter tubular support for the scuttle, which was doubly cross-braced. In addition there was a multiplicity of smaller diameter tubing and fabricated bracketry to carry the body structure and mechanical components.

In bare form the whole construction looked more like a product from Italy than from the outskirts of London. The bodywork enhanced this illusion, and in fact the prototype bore an uncanny resemblance to the Ferrari *Barchetta* of the period. By the time the car went on sale the lines of the body had been slightly altered and the nose restyled, but the influence was still apparent and the manner of its construction – hand-formed in aluminium over the tubular frame and attached by pop rivets – was pure Italian sports racing car practice.

This was the car, a most unusual product for an English manufacturer, that went on sale after its launch at the 1953 Motor Show. Necessarily fairly expensive at £950 owing to being hand-built, the Ace was still nearly £200 cheaper than the successful Jaguar XK120. Admittedly the latter had its wonderful 3.4-litre engine which gave it 120mph performance, but only a brave or foolhardy Jaguar owner could have shown the Ace a clean pair of heels on an English country road. The Jaguar's plebian chassis – with its origins in the bulky MkV saloon – and indifferent brakes frankly handicapped its performance on the sinuous roads with which England abounded at the time, but, in fairness to William Lyons, the car was primarily intended for the American market.

The new AC received an excellent response at the show and 1954 began with 23 cars sold, three of these going to the US. In this context one must remember that the Ace was a hand-built car that was not suitable for – nor intended for – large-scale production.

A coupé version of the Ace was introduced at the 1954 Motor Show and called the Aceca, reviving a name that the company had first used during the 1920s. The new car, if anything, received an even more enthusiastic

BEGINNINGS

The very first Cobra, initially painted metallic yellow, was in constant demand as the only demonstrator and show car, but here Shelby himself is giving it a gentle airing along the Californian coast some time in mid-1962.

reception than the Ace, and the press were particularly complimentary about it. During 1955 the factory produced 62 Aces and 45 Acecas, 13 and seven respectively going to the US.

The Ace had been entered for rallies and a few races, with some success. Exemplary though the car's handling and brakes may have been, however, the AC engine, by now over 30 years old, had never been designed with racing in mind. Sussex-based AC dealer Ken Rudd, remembering the success that one of the original Tojeiro sports racers had had when fitted with a Bristol 2-litre engine, and also bearing in mind the Cooper Bristol single-seaters and Frazer Nash sports cars, decided in 1955 to fit one of these power units to his Ace. As a result Rudd won the Production Sports Car Championship in 1956, by which time AC itself had introduced the Ace Bristol as a higher-performance alternative to the AC-engined Ace. Even in its first year of production the new model outsold the original Ace, with sales of 62 and 44 respectively. US customers took 34 AC-engined and 30 Bristol-engined examples.

The following year, 1957, saw the greatest number of Ace sales, with the Bristol-engined version way ahead at 155 cars (97 for the US) against 21. The little English sports car with the svelte looks, raucous exhaust note and really impressive performance 'straight out of the box' had arrived! That year it rewarded its followers in the US with third place at the Sebring 12 Hours endurance race in the sports car class, behind a pair of Ferraris. And, in addition to numerous victories and placings at less important events, it won the Sports Car Club of America Class E Championship. Back in Europe Rudd entered the Le Mans 24 Hours, finishing second in the 2-litre class and 10th overall.

Among the cars shipped to America that year, one – chassis number BEX 327 – was specifically requested without an engine or gearbox. Could this have been a portent of things to come?

Carroll Shelby, in the meantime, had carved out for himself something of a reputation as a racing driver and in 1956 had won the big car class of the SCCA Championship driving a 4.9-litre Ferrari. He may not have noticed, but an Ace Bristol won its class in the last race of the series that year at Palm Springs.

During 1957, 1958 and 1959 the Ace Bristol completely dominated SCCA Class E races and as a result, in an attempt to give other competitors more of a chance, it was moved up to Class D for 1960 and then

9

The definitive Ace/Cobra shape was particularly apparent from this angle, showing a 1962 RS 2.6 Ace with curved windscreen.

Class C for 1961. It continued winning – and so did Shelby. His finest hour, in Europe at any rate, was victory at Le Mans in 1959, partnered by Roy Salvadori in a works Aston Martin DBR1. The winner of the 2-litre class, in seventh place overall, was the indomitable Ken Rudd with Peter Bolton in an Ace Bristol.

Aston Martin discontinued its racing programme in 1960 and Shelby, who had been a works driver for some time but was beginning to suffer the effects of a heart condition, decided to retire from active participation as well. He was 37 years old. He found plenty to do back home, however, for he became west coast distributor for Goodyear racing tyres, a consultant to *Sports Car Graphic* magazine, and started the Shelby High Performance Driving School at the Riverside race track.

Besides the realities of everyday life we all have our day-dreams and Shelby was no exception. His constantly recurring one was of a sports car that would combine the handling and roadability of European products with the characteristics of a powerful American V8 engine.

The choice of a Chevrolet V8 seemed the natural one but, when approached, the hierarchy at General Motors were unenthusiastic. They already had their own sports car in the form of the Corvette and judged that it would not be very productive to equip a possible competitor with the same engine. Shelby considered the Austin-Healey 3000 as a recipient for an American engine but also met with little enthusiasm from that quarter. In retrospect that response was in all probability most fortunate if one considers the result of a V8 with over 250bhp propelling the Healey's rather primitive box-section tin chassis with an assemblage of Austin saloon bits and pieces hung onto it, not forgetting the increase in the already unbearable cockpit heat generated by the car's existing 3-litre, six-cylinder engine.

It was Ford who unwittingly provided the answer to the engine problem by supplementing its ageing Y-block V8 engines with a fresh V8 design, which employed a new thinwall casting technique that resulted in an extremely light and compact motor. It was of conventional pushrod overhead valve configuration but had, for the time, an extremely short stroke. The initial capacity was 221cu in (3.6 litres) but allowances were made for this to be increased.

Shelby had a passing acquaintance with the head of Ford's stock car department, one David Evans, and to him he outlined his ideas for a sports car together with a request for a couple of the new engines in order that he

Within a month of its completion, the first right-hand drive Cobra – taken from the initial batch of cars for the US – made its public debut in Europe at Earls Court in October 1962. This is CS 2030, painted red with black trim.

could evaluate their suitability. Evans, aware of Shelby's reputation in the racing world, took him seriously enough to comply. To provide storage and office facilities for his tyre operation, Shelby had rented part of the building from which his good friend Dean Moon operated his tuning equipment business. It was here that the Ford engines arrived, and the two men quickly realised that this basic motor had much potential and could suit Shelby's intended project very well. But what about a car to put it in?

Over in England the makers of the Ace Bristol had just been informed by their engine suppliers, Bristol Cars Ltd, that in the future all Bristols were to be fitted with Chrysler V8 engines. As a result Bristol was going to discontinue the manufacture of the BMW-based six-cylinder engine, which the company had been happy to supply to firms such as AC, Frazer Nash and Arnolt as well as fitting to its own products. There were enough Bristol engines to be going on with for the moment, but in any case sales of the Ace had fallen off quite severely with only one AC-engined and ten Bristol-engined cars sold during 1961, the only one destined for the US being the AC-engined example.

The AC factory had for years relied on other sources of income besides making sports cars, its output ranging from golf trolleys to invalid carriages. But the downturn in orders, coupled with the imminent discontinuation of Bristol engines and the decision to have to cease manufacture of its own long-running 2-litre engine, saw the company at a crossroads. Once again Ken Rudd came into the picture with a suggestion that the Ace could be given another lease of life by fitting the 2.6-litre Ford Zephyr engine. The prototype was a re-engined normal Ace, but when the small run of these cars began to be manufactured they not only had the heavier gauge main chassis tubing hitherto reserved for the Aceca coupé but the bodywork was also restyled. The short-stroke Ford engine allowed the bonnet line to be lowered and the air intake reduced in size, resulting in a leaner and more aggressive look.

Although he was thousands of miles away from the machinations of the British motor industry, Carroll Shelby was kept abreast of anything that was of interest to him through his consultancy at *Sports Car Graphic*. And so when he heard about the various goings-on at the AC factory he remembered the neat little roadsters that had had so much success in production sports car racing over the previous few years. The new-look Ace certainly brought the car into the fresh decade. And if a six-cylinder Ford engine could be fitted, why not a V8?

In almost no time at all Shelby was in touch with the Hurlocks, who had owned AC Cars Ltd for the previous 30 years, and made arrangements to visit them. The response was very different to some others that he had experienced while in pursuit of his dream, and it all looked very encouraging. Upon his return to the US he found his contacts at Ford still sufficiently interested in his project to supply a couple more engines, but this time they went directly to the AC factory in England. At the same time the design staff there got busy with any modifications thought necessary to the 2.6 Ace so that it could accommodate the larger engine. On early drawings the car even received a revised designation – the 3.6 Ace – but in the event this title was rendered prematurely redundant due to Ford introducing the larger 260cu in (4.2 litres) version of the motor that was to power the first production cars.

ESSENTIAL AC COBRA

The basic chassis frame remained unaltered apart from the addition of an extra tubular cross-member just forward of the differential, but the rear suspension tower was considerably modified so that the differential could be rubber-mounted as a complete unit within the structure rather than built-in as on the Ace. A good deal of the bracketry around this area was also reinforced, the rear uprights were cast rather than fabricated, and the drive-shafts and rear hubs were more substantial. The first car was built with inboard brakes but this idea was subsequently abandoned.

Within a few days of the unpublicised Silverstone test, after Carroll Shelby had had time to discuss its outcome with those involved, it was decided definitely to go ahead with the project and contracts were drawn up. The AC company would produce the complete car minus engine and gearbox in batches as required, and the AC identity was to remain. Ford in turn would supply the engine and gearbox in return for which, apart from payment of course, there would be a discreet 'Powered by Ford' badge on each of the car's flanks. Thus Shelby was able to realise his dream with almost no finance and no lengthy – and costly – development.

The first car was chassis number CSX 2000. This car has sometimes been erroneously given the chassis number CSX 0001 in publications, and furthermore the CSX prefix has been said to indicate 'Carroll Shelby Experimental' whereas in reality it stands for 'Carroll Shelby Export'. It had its engine removed and was sent out from the AC factory, unpainted, on 20 February 1962 bound for New York by air and consigned to 'Carol' Shelby (sic). Under the heading 'description of car' in the factory ledger it was described, as were all leaf-sprung Cobras, as an 'AC Ace Cobra', which dispels another myth that the car arrived nameless in America and Shelby dreamed one night that he saw 'Cobra' written on the nose and thereby arrived at its title.

When it landed in the US the car was transported to Dean Moon's speed shop in California where a 260cu in (4.2-litre) engine awaited it. This engine, XHP 260 1, was the first of the 'High Performance' motors that were supplied to Shelby in both 260 and 289 (4.7-litre) form by Ford, and were fitted into both Cobras and Shelby Mustangs. These engines were more robust than normal

What price for these bits and pieces now?

and had, among other features, solid tappets rather than hydraulic ones.

Legend has it that by the evening the car was up and running, and for the next few weeks provided Shelby's friends and other selected disciples with a taste of what was to come. What is certain, however, is that Shelby, anxious from the start to make the car his own, disregarded his agreement with AC Cars Ltd, removing the English company's badge from the nose and having Shelby painted in script thereon before the car was even sprayed. Admittedly he then applied a rather crude cast badge carrying the words 'Shelby Cobra' so prominently that they dwarfed the tiny AC motif in its centre.

Easter was the time of the annual New York Motor Show. After Shelby had demonstrated the prototype Cobra to his friend Dave Evans, Ford invited him to bring it to be exhibited on the company's stand. Hurriedly the car was painted metallic yellow and detailed before being delivered to the show. The little sports car, although dwarfed by behemoths from Ford and other manufacturers, almost stole the show – and that was from a static display without the onlookers having sampled its performance! Shelby returned from the show elated, his pocket book full of enquiries and orders from both dealers and public. It was to be some time before the next Cobra was despatched from AC, however, and until that time car number one had to fulfil myriad tasks, from press and customer demonstrator to development hack.

At length, on 19 July, car number 2 (chassis number CSX 2001) left the factory in England bound for New York by air. It was followed, before the end of the month, by the next five cars, the first being flown direct to Los Angeles and the others going by boat to New York. CSX 2002, painted red and trimmed in black leather before it left the AC factory, became the first racing car, as described later in this book (see page 26). So it was that the Cobra project gradually came to fruition and a legend was born…

LEAF-SPRUNG COBRAS

July 1962 saw the first of the customer Cobras leave the AC factory bound either for the east or west coasts of the US, depending on where they were to have their engines and other bits and pieces – such as instruments – fitted. Partially to spread the work-load but also to avoid unnecessary transportation, a number of cars were sent via New York to Ed Hugas's dealership in Pittsburg to be finished. Those that were shipped to New York mostly went by sea, but the greater number of cars went to Los Angeles and were usually air-freighted. By the end of 1962 a total of 61 cars had arrived in the US.

These were all the 260cu in (4.2-litre) variety, almost invariably referred to retrospectively – and colloquially – as the 'MkI'. They were little changed from the first car other than that the inboard rear disc brakes were now of conventional outboard type. With a diameter of 10⅜in, in conjunction with larger 11¹¹⁄₁₆in discs at the front, these

An as yet unregistered right-hand drive 289 roadster in white (top), one of six European cars sold in this colour, being driven out of AC's Thames Ditton showroom. In all probability the first Cobra (above), now in bare aluminium and equipped with a roll-over bar, testing while emblazoned with 'Powered by Ford'.

13

Colour advertisement comes from *Road & Track* dated May 1964. Cobras ended up being used for all manner of publicity and 'other' photography, but what red-blooded male could resist a ride with this lady?

give the Cobra very respectable braking even though the system is not servo-assisted and consequently needs a determined right foot.

In September 1962 *Road & Track* published a road test of the first Cobra. The tone was generally praise-worthy but the magazine did comment that a fair degree of effort was required on the steering wheel to force the car into corners as speed built up. For some reason the top speed and acceleration of this car was far superior to a standard 289 (4.7-litre) version tested by the same magazine two years later, but, whether or not this was a 'trick' car with special performance, the figures make fairly astounding reading, especially when you consider that this exercise took place 35 years ago.

The 0-60mph sprint occupied 4.2sec and 0-80mph took 6.8sec, while the magic 100mph came up in an astonishing 10.8sec. The highest recorded speed that day was 153mph at 7000rpm, with the tester relieved to find that the brakes – 'the best we've ever tried' – were up to all this. The price of this incredible machine was around $6000, depending on specification, but with such things as a heater and seat belts listed as extras.

On 21 January 1963 Cobra chassis number CSX 2075, painted black with a black interior, left the AC factory and was put aboard *SS Dohedyk*, along with eight other Cobras, bound for Los Angeles. When it arrived at the Shelby factory it was the first car to receive the 289cu in (4.7-litre) motor and from then on all leaf-sprung cars were so equipped.

It was not until chassis number CSX 2126, completed in January 1963, that the steering box hitherto used on

Chassis number CSX 2042 (facing page, top) was originally supplied in November 1962 as a standard 260 but was club raced in the US from new and uprated to 289 power after its first season. Early cars have no bonnet handles or side vents on the front wings, but this one has an added bonnet scoop – as used on works cars – and roll-over bar. The engine bay shows the pop-riveted insert panel on the inner wing to give clearance for the dynamo on early cars; at the time the rocker covers would have been steel but this 289 motor has been dressed up a little. The 100th Cobra built (bottom) is a slightly later car, combining factory-fitted 289 engine with a steering box, and survives in remarkably original condition.

ESSENTIAL AC COBRA

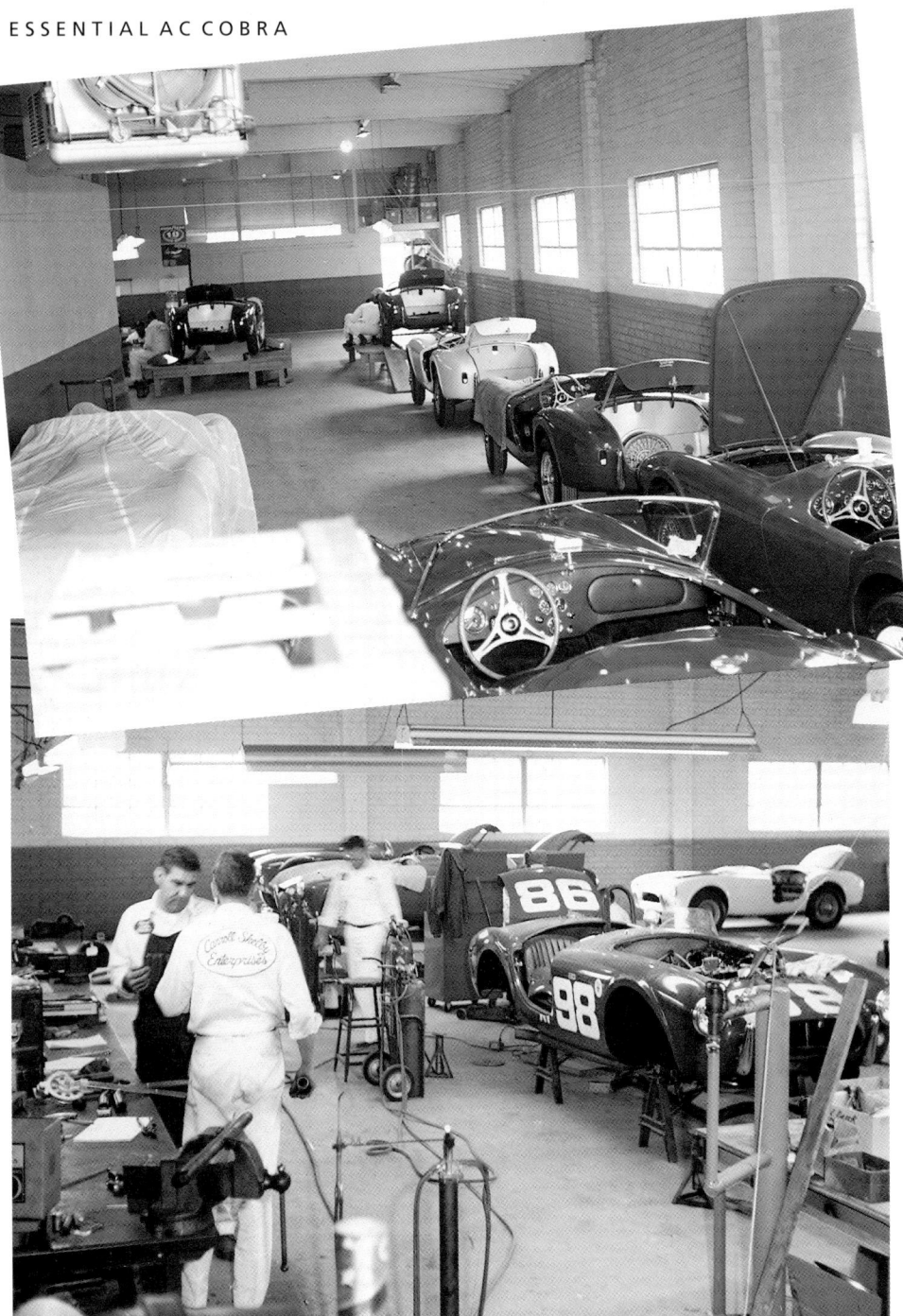

Early Cobras at the Carroll Shelby Enterprises workshops in Princetown Drive, waiting to have their engines fitted. All pre-date chassis number CSX 2126, so they have Ace-type steering and distinctive, flat, wood-rim wheel. View 'round the corner' shows the car that Billy Krause drove at Riverside in October 1962 – the Cobra's racing debut.

the Cobra and the Ace before that was replaced by a rack and pinion set-up. Although the Bishop Cam steering had proved perfectly satisfactory on the Ace, the wider tyres fitted to the Cobra did have a greater tendency to pick up road irregularities and transmit these to the steering, so the rack and pinion system was a great improvement. As a matter of interest, the very first Ace chassis shown at the 1953 London Motor Show had also been equipped with rack and pinion steering.

Shortly after this, during March 1963, it was necessary to build a couple of right-hand drive cars for racing. Before construction of the European series of right-hand drive and left-hand drive Cobras began, therefore, two Cobras bearing chassis numbers 2130 and 2131 were built with a CS chassis prefix, denoting that they were right-hand drive cars for the home market. Following on from CS 2030, the very first right-hand drive car built to serve as a demonstrator for the factory, these were the first cars built by the AC factory specifically for racing, as opposed to cars converted by Shelby for this purpose once they had reached America.

The Cobra made with the 289 (4.7-litre) motor is often referred to as the 'MkII', but it was in fact merely a development of the leaf-sprung car with first the larger engine then later the rack and pinion steering. The factory never called these successive versions 'MkI' and 'MkII', and, to the best of my knowledge, neither did

LEAF-SPRUNG COBRAS

This beautifully restored 289 is a late car, chassis number CSX 2503, which left the AC factory during July 1964. Black with red interior was not one of the most common colour schemes, but one of the most attractive. Once production had settled down from chassis number CSX 2201, this was the standard instrumentation for US cars, with dials by Stewart-Warner except for the futuristic clock, which came from a Ford Galaxie. Steering wheel boss on CSX cars could either have a traditional AC badge, as here, or one with a Cobra motif.

17

Period Shelby publicity shot of a standard 289 roadster with whitewall tyres beside one of the FIA cars with wide rear wings and cut-back doors – racing Cobras frequently used the number 98. Rear view shows the same standard 289 with weather equipment in place; although much maligned by the press, especially in the US, the top is both good-looking and practical.

Carroll Shelby. It was, by the way, around this time that it dawned on the AC factory that it had been spelling Shelby's first name incorrectly. The first attempt to rectify this in the factory ledger was at chassis number CSX 2120, where he became 'Caroll' Shelby, but it was not until CSX 2143 that they finally managed to get the 'Carroll' right.

Once the Cobra was standardised with the new steering and larger engine the package seemed about right, and the AC factory sent a sustained flow to America as the cars were built. The vast majority went to the Shelby plant at 1042 Princetown Drive, Venice, California, with only 14 out of the 460 leaf-sprung 289s built consigned to the East Coast. The majority of the racing versions were built in the same chassis series. Just after the first two-right hand drive cars, therefore, three Le Mans replicas (CSX 2136, CSX 2137 and CSX 2138) were built, followed by the left-hand drive Le Mans car (CSX 2142) for Ed Hugas – and even the chassis for the racing Daytona coupés were taken from this sequence.

LEAF-SPRUNG COBRAS

Two photographs and accompanying publicity blurb put out when Shelby's 'Cobra Caravan' **was touring the US. Miss Mernone is actually seen, of course, in a 289, not the 427 promised.**

```
C O B R A   C A R A V A N

            WHO SAYS IT'S A MAN'S WORLD?

     Miss Pat Mernone, 24, of Washington, D.C., is an accomplished
graduate of the Carroll Shelby School of High Performance Driving who
has abandoned the sports car circuit for the winter to accompany the
Cobra Caravan in its cross-country tour of the nation.  Miss Mernone
is driving street versions of Shelby's 427 Cobra and 1966 GT 350
Mustang on the 5,000 mile journey and providing demonstration rides
for press media representatives.  She will also appear with the display
cars at each of the Caravan's 12 stops.

Shelby American, Inc., 6501 W. Imperial Hwy., Los Angeles 9, Calif.
```

As supplied by Ford, the standard 289 High Performance motor fitted to road-going Cobras had a compression ratio of 11.6:1 and was quoted as producing 271bhp at 6000rpm, with maximum torque of around 312lb ft at 3400rpm – and all this in a car that weighed a little less than a ton. Even the 1964 *Road & Track* road test, which generated a considerably slower but probably more realistic set of figures than the magazine's previous exercise with the first Cobra in 1962, gave a top speed of 139mph and a standing-start quarter-mile time of 14.0sec. Petrol consumption for this particular car, fitted with a Ford Autolite four-barrel carburettor, was quoted at 13-18 miles per (US) gallon.

The gearbox employed on all the leaf-sprung Cobras, apart from some very early 260 versions which used a Ford unit, was the Borg Warner T10.

Bodywork on standard road cars came in two varieties. All 260s and 289s until chassis number CSX 2160 had small, flat-sided wheelarch flares and no side vents on the front wings. The very earliest cars were fitted with proprietary overriders as used on Rovers, Aston Martins and early 2.6 Aces, before AC adopted the specially manufactured overriders used on later 2.6 Aces and all subsequent Cobras.

With the increase in wheel rim width from 5½in to 6in after chassis number CSX 2160, the wheelarch flares

19

Under the skin of the July 1964 Cobra 289. Alternators had been fitted to US cars since chassis number CSX 2201; transverse leaf springs at front and rear were bound with sticky brown tape, and inner wing aperture through which front spring passes had rubber mat roughly cut to shape and pop-riveted in place. Boot view shows utilitarian finish and spare wheel well in moulded white glass-fibre.

were slightly enlarged and at about the same time side vents began to be fitted into the body sides behind the front wheelarches. The bonnet was at first secured by a pair of budget locks, as had been the case on all Aces, but after a while chromium handles were employed, to avoid fiddling with the key.

Instruments on the very early cars were similar to those on the 2.6 Ace with the exception of the rev counter, which was an American component. Later a 4in (instead of 5in) speedometer was used, normally in conjunction with a Rotunda rev counter; a Sun rev counter was also fitted to some cars between chassis numbers CSX 2130 and CSX 2200. From then on Stewart-Warner gauges were used on the Cobra, the only exception being the clock – a rather horrible thing sourced from a Ford Galaxie.

LEAF-SPRUNG COBRAS

Colour cover of AC's rudimentary four-page brochure put together when sales of right-hand drive leaf-sprung 289s began in 1964.

So what was this snorting, fire-breathing monster like to drive? In reality it is a soft, forgiving car that can be propelled with no fuss, no wheelspin and no opposite-lock, its large, lazy V8 engine wafting you towards a higher gear. The effortless performance may mean that you are travelling more quickly than you imagine and bring you into bends and other obstacles deceptively fast, but reach for the brakes and treat the car with respect and it will not let you down. It is sure-footed, well-braked and – due to its excess of power – amazingly safe. Play the hooligan with it, however, and – watch out! – you had better be a good driver. Otherwise you will, at best, frighten yourself into near incontinence or, at worst, leave the road in an ungainly fashion.

A legend the Cobra may now be, but in their time these cars did not sell in great numbers and the AC factory was more or less able to keep up with demand. Towards the end of 1964 Carroll Shelby was anxious to get on with the MkII version of the Cobra and the last American-specification leaf-sprung Cobra roadster, chassis number CSX 2589, left the AC factory on 20 November. It was painted rouge iris and had a black interior – a popular colour scheme. Together with the preceding five roadsters, it was put aboard the *SS Pacific Fortune*, bound for Los Angeles. Including FIA-bodied versions, a total of 583 leaf-sprung Cobra roadsters were produced in this form during the model's three short years of production.

European leaf-sprung Cobras............

Production of Cobras for the UK and European markets began a good while after the US series of cars. There was, however, one 260 version from the US series, chassis

21

ESSENTIAL AC COBRA

Back cover of UK sales brochure shows optional glass-fibre hard-top, while UK launch advert promises right-hand drive Cobras from September 1964.

number CS 2030, completed on 9 September 1962 and painted red with black leather trim, made in right-hand drive form and retained by the AC factory in Thames Ditton for demonstration purposes.

It was not until 1 October 1963 that the first of the European series left the factory, this one, chassis number COX 6001, going out to the French distributor, Chardonnet, for the Paris Salon. In fact the second car made, COX 6002, was booked out of the factory a day earlier, on 30 September, bound also for Chardonnet and the Paris Salon. Chardonnet must have sold at least one more car as a result of this motor show appearance by these two cars, as the next Cobra in the series also went to him on 26 November.

The chassis numbering sequence was completely different for the European cars. Whereas the US series used the prefix CSX (only the handful of right-hand drive cars had a CS prefix) and were numbered from 2000, the European series were prefixed COX for left-hand drive and COB for right-hand drive, and furthermore they were numbered from 6000.

The first right-hand drive Cobra made, apart from the aforementioned examples bearing a CS prefix, was COB 6004. This car has been owned for many years by Lord Cross, who bought it when almost new and since then has made the fullest use of it imaginable, having driven it regularly on the road as well as racing it.

22

LEAF-SPRUNG COBRAS

Autocar road test of CPO 681B in November 1965 recorded 0-60mph in 5.5sec, standing-start quarter-mile in 13.9sec and a top speed of 138mph. The exercise was not carried out along this stretch of road in leafy suburbia...

Since they did not appear until towards the end of 1963, all European Cobras – with the exception of the CS 2030 demonstration car plucked from the US series – are of the 289 variety with rack and pinion steering. Perversely, the car which the factory designated as the prototype for this series, COB 6005, was not the first in the chassis number sequence. It was retained by the factory and not sold until November 1965.

Considering the legend that has built up around the Cobra since its inception nearly 35 years ago, it has always surprised me that so few cars were actually sold in Europe when new. Admittedly during 1964, the year in which the largest number of leaf-sprung Cobras of all types were produced, 31 cars were sold in the UK, but Europe as a whole took only six – four to Switzerland and two to France. In latter years many times that number must have returned across the Atlantic…

Apart from their Smiths instrumentation and, of course, right-hand drive where applicable, European cars were to the same specification as their US counterparts with the exception of the badge on the boot lid. Whereas the CSX cars have a Cobra badge, the COB and COX cars have a large, die-stamped, chromed AC motif.

Autocar magazine of 12 November 1965 published a road test of Ken Rudd's personal Cobra and, in the typically understated manner of that periodical at that time, the verdict was mildly enthusiastic. The tester managed to extract a top speed of 138mph and obtained acceleration times of 0-60mph in 5.5sec, 0-80mph in 8.9sec, 0-100mph in 14.0sec and the standing-start quarter-mile in 13.9sec. All these figures were slightly better than *Road & Track* had recorded in its 1964 road test of a similar 289, except for the top speed being 1mph slower. *Autocar* achieved an overall petrol consumption of 15.1mpg and gave the list price as £2732 4s 9d (including purchase tax).

Cobra's beauty with lean body style of leaf-sprung models is still apparent in this unusual view of chassis number COB 6020, built in November 1964.

The magazine remarked that the good grip provided by the independent rear suspension resulted in only slight wheelspin in a straight line during the timed tests. The writer praised the way the car could be powered through corners with the tail out, but commended a good deal of restraint in this department in wet weather. The brakes were powerful but the tester noticed, not surprisingly, that high pedal pressures were needed. Lack of refinement in areas such as heating and demisting, the soft-top and the absence of exterior door handles (the car cannot be locked) did receive adverse comment, but the test was summed up as follows.

'There can be no denying that the Cobra is a very exciting car to drive, or even passenger in. The acceleration is sensational and very similar to taking off in a piston-engined aircraft with open cockpit. In much the same way, nearly all creature comforts must be sacrificed for performance – a condition which is worth it most of the time, but very occasionally not. It is a fine-weather car for clear skies, open roads and a life away from it all.

Part of its sorcery lies in its ability to instil the same exhilaration from a short run up the road on a Sunday morning, but most of it comes from that aggressive thrust that is always more than enough for any situation.'

The tester almost had it right, but some of my more religious moments with Cobras have been with the soft-top stowed in the boot on a crisp, starry night, out on a switchback English country road with the engine cogged up in a high gear, the exhaust note warbling and blatting back from the hedgerows. Or, as a contrast, top down in a hurry across the Wiltshire Downs early on a Sunday morning, driving rain and a soaking road, wipers flailing and an occasional wipe of the 'shammy' to clear the interior of the 'screen, going too fast, maybe, on a road that could only appear almost deserted. The occasional dawdling car in a dip, the lithe beast twitching as it crosses

Smiths instruments were always fitted to cars with COB or COX chassis number prefixes. Bonnet handles of the correct type have these escutcheons; on earlier Cobras the bonnet was secured by budget locks released by a T-handle. Two of the early badge styles: the 'Shelby AC Cobra' version was fitted before chassis number CSX 2054, the 'Cobra' one to some subsequent CSX and first-series COB/COX cars; a revised version of the 'Cobra' type was found on later 289s and all 427s (see page 66).

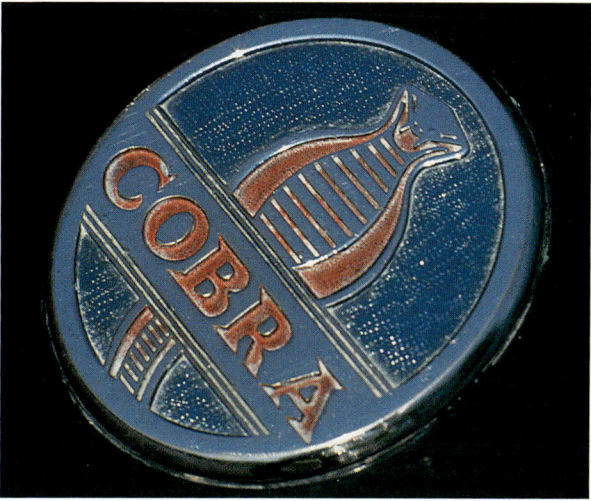

to overtake under power, and that delicious, selfish satisfaction when you reach the destination wind-blown and damp, noise ringing in your ears, knowing that you have been dangerously foolish but come out on the other side.

The European leaf-sprung Cobra outlived its American counterpart – that ceased production at the end of 1964 – with 15 cars leaving the factory during 1965 and two more trickling out in 1966, one in February (COB 6054) and the other in July (COB 6055). These two, for some reason, were out of sequence as the final chassis number of the European series was COX 6062, which left the factory on 3 May 1965, painted bright blue with grey trim, bound for Swiss agent Hubert Patthey. In total just 60 European cars, with COB and COX chassis numbers, were built in this series.

RACING IN THE USA

With the pedigree of the Ace, a great deal more power and manufactured under the auspices of one such as Carroll Shelby, it was almost inevitable that the Cobra should end up on the race track. Even before cars had begun to arrive from England in sufficient quantities for sales to begin, one of the first Cobras, chassis number CSX 2002, was prepared for motor racing.

Much of the work was done by Phil Remington, who had come to the Shelby organisation when it had taken over the old Scarab works. Fairly minimal attention in the mechanical department was restricted to fitting a Spalding 'Flamethrower' magneto and removing the air cleaner. The bodywork was modified to include air scoops for the front brakes, two smaller subsidiary ducts under the main intake to aid engine cooling, and a small air scoop on the bonnet coupled with five rows of louvres at its rear end to improve engine ventilation. The full windscreen was replaced by a vestigial Perspex one, the obligatory roll-over bar (insisted upon by the SCCA) was fitted, and the racing tyres were Goodyears.

Phil Hill (top) during the 1963 Sebring 12 Hours in the best-placed of five works Cobra entries, with third in the GT class and 11th overall. Tactical thinking by Shelby caused drivers to be shuffled around mid-race: **Ken Miles took over car 12 instead of nominated co-driver Lew Spencer, while Hill relieved Dan Gurney (above) in car 15 after its two-hour delay for steering and gearbox repairs.**

One of the Carroll Shelby Enterprises race shop employees, Ralph Falconer, working on a 289 motor.

A good deal of testing was carried out during the late summer of 1962 by Pete Brock, who had originally joined Shelby to teach at the High Performance Driving School but transferred to the Cobra project when it became a reality. For the car's first race Brock was rather disappointed not to be given the drive, but Shelby felt that, excellent driver though Brock was, he lacked the essential competitive edge and the seat went to Billy Krause, an up-and-coming semi-professional driver.

The event was a three-hour endurance race run as a preliminary to the *Los Angeles Times* Grand Prix at Riverside on 13 October 1962. That the Cobra was allowed to race at all was an oddity as, under the rules of SCCA production classes, a certain number of cars had to have been built, and yet this was only the third Cobra in existence. A dispensation was granted to Shelby, however, because Chevrolet also wished to enter the new Corvette Sting Ray (with 327 motor) that had only been launched the previous month, so both firms were allowed to run – to make up the numbers and provide interest – but not to compete for championship points.

The start of the race saw the solitary Cobra mixing it at the front with a shoal of Sting Rays, but by the end of the first hour it was in the lead by something over a mile. Shelby and his small band must have been ecstatic but then, to the relief of the Sting Ray drivers gradually dwindling into the distance, a rear hub carrier failed. This left Doug Hooper to win, followed home by the other Sting Rays. Shelby, though, had tested the might of General Motors and won a moral victory. The sight of his little roadster running right away from the monster Corvette Sting Rays would not be forgotten. Admittedly the Cobra was a lighter and more specialised car, but its reputation was now in the making.

One more race, or rather series of races, was contested that year – the Nassau Speed Week held during December. This somewhat strange event in the Bahamas was the brainchild of the characterful Captain Crise and was run on an appalling track sketched out at the old Oakes airport on the outskirts of the capital. The event was well-known as a car-breaker, but it was also convenient as a test of durability. Victory added no points to any championship, but the Nassau Speed Week had become something of a mecca for those involved with racing – they loved this winter break in the sun! – and attracted a good deal of publicity. Shelby was no doubt also attracted by free entry, transportation and accommodation, and two cars – CSX 2002 and CSX 2011 – were sent. The original car was entered by Shelby and the second one by Holman & Moody, Ford's stock car department run by Shelby's benefactor Dave Evans. Although this second car was also prepared by Shelby, it was less specialised, lacking the bonnet scoops and louvres and having just the extra brake ducts fitted at the front. Among other improvements since the Cobra's racing debut some two months earlier, Phil Remington had ensured that the rear hubs were strengthened. Hopes were high provided the cars could survive the dreadful pounding meted out by the track surface…

In the first race for GT cars Krause again ran away from the rest until something broke, this time part of the steering mechanism. The Holman & Moody car, driven by John Eberly, was not up to the same speed and ran only on the first two days before the scanty Shelby staff gave up trying to keep two disintegrating cars raceworthy and concentrated on the fastest. By the time of the last event, the three-hour Trophy Race for prototypes and GTs, chief mechanic Don Pike reckoned the Cobra might reach the finish as he had spent the previous few days re-attaching and reinforcing broken pieces, and beefing up anything else that might break.

In this free-for-all race Krause amazingly managed to stay with the front-running Ferraris and other pure sports racers, leaving all his GT rivals way behind. After one and a half hours he was running very strongly but the twin tanks, fitted one above the other specially for this event, were by now almost empty and he was called in for fuel. After Carroll Shelby himself had carried out the refuelling with several cans and Don Pike had made a couple of adjustments under the bonnet, Krause was hurriedly sent on his way without even losing a place. Three-quarters of an hour later Krause had actually begun to improve his position and could even have been in contention for overall victory, but then the car failed to reappear past the pits. It had run out of petrol on the far side of the track. Poor Shelby was very honest and admitted that the fault was most probably his: in his enthusiasm he had not allowed time for the hastily-poured petrol to overcome the air lock between the two tanks, so the car had been sent out only partially filled.

Winner or not, the Cobra had made a storming appearance. Shelby returned home with the car's reputation further enhanced and with more than a few dealership enquiries and orders.

Shortly after this Krause was wooed away by Chevrolet, with the offer of more money, to become a Corvette Sting Ray driver. In the event Chevrolet aborted its racing programme and he was left without a drive, his seat in the Cobra meanwhile having been taken by a young driver called Dave MacDonald, who was rapidly making a name for himself with his wild-looking but effective driving style. Almost immediately the team brought in a second driver, the wily expatriate Englishman Ken Miles, who could also contribute enormous engineering and development experience.

January 1963 saw the Cobra's first race win, in a minor event at Riverside. MacDonald was closely followed by Ken Miles, who had amused himself during the race by unnecessarily calling into the pits for a drink of water and then resuming behind the Corvettes, only to pass them all and close up behind MacDonald again…

With greater things in mind, however, Shelby took the bull by the horns and entered three cars for the 1963 Daytona Continental, one of the first events on that year's FIA GT calendar. The drivers were to be Dave MacDonald, Skip Hudson and Dan Gurney, but even with the services of the latter an outright win was out of the question. This was a completely different ball game from anything attempted so far, the opposition including cars such as the omnipotent Ferrari 250GTO.

The Shelby team probably surprised themselves in the three-hour race with outsider Hudson's Cobra swopping the lead with Pedro Rodriguez's Ferrari until the V8's flywheel burst on the 54th lap, injuring Hudson and causing him to crash. Gurney, who had started at the back of the field after a last-minute engine change, charged through the opposition towards the lead but suffered total ignition failure, so it was left to MacDonald to bring the only surviving Cobra to the finish, in fourth place.

The 260 engine was soon to be superseded by the 289 and to try this out two cars were entered for an unimportant local SCCA event in which Ken Miles and Dave MacDonald came first and second.

The Sebring 12 Hours was the next FIA event that the Cobras attempted and five cars were prepared, four entered by Shelby and one by Holman & Moody. The main team cars were to be driven by MacDonald/Roberts, Gurney/Hill and Miles/Spencer, the other two cars running in more standard form. It was not a good race for the Cobras. After a series of problems the Miles/Spencer car finally fell by the wayside with a detached steering rack, while MacDonald/Roberts retired when a rear end oil seal failed. The Gurney/Hill car suffered brake overheating and electrical problems but finally finished 11th overall and third in the big GT category.

The SCCA had just inaugurated the USRRC (United States Road Racing Championship) and so this, as well as the SCCA Class A Championship, was contested. The three factory cars, with Bob Holbert now in the team as well, were to concentrate on USRRC events while privateers would try for the Class A production races, occasionally supported by the works drivers. The privateeeers – Bob Johnson, Dan Gerber, Ed Leslie and Tom Payne – were wealthy businessmen and three of them had previously raced Corvettes.

Bearing in mind the success of the Ace Bristol in US production racing, the Cobras should have done very

LEARN TO RACE! The controlled limit of adhesion is yours after one week at Riverside International Raceway with the Instructors of the Carroll Shelby School of High Performance Driving. We can teach you more in a few days than you can learn by yourself in one season! Send One Dollar for illustrated Brochure to: **CARROLL SHELBY SCHOOLS, INC. DEPT. A, 6501 W. IMPERIAL HWY. LOS ANGELES, CALIFORNIA 90009**

RACING IN THE USA

Bob Johnson won the 1963 SCCA Class A Production title. Later in the season this car was modified with side vents in the front wings, as occurred on standard Cobras.

well, but at first this was not the case. The team cars suffered from overheating rear axles at their first race in the series at Pensacola in Florida and the amateurs had little more luck, but then Johnson had a breakthrough with two wins in a row at Cumberland and Bridgehampton. At Laguna Seca the team cars once again ended up with seriously overheating rear axles, although MacDonald won a qualifying race. Prepared for this eventuality, however, the mechanics fitted the cars with differential oil coolers for the finals – and they finished first, second and fourth. From then on the Cobras, both team and private, were virtually unbeatable, even when they occasionally came across the very special Chevrolet Sting Ray Grand Sports that sometimes ran in the modified classes, and by the end of the season they had captured both the USRRC constructors' and drivers' titles as well as the SCCA Class A series.

To round off the season the Bridgehampton 500kms FIA race was contested in September by the works team, with Holbert, Gurney and Miles as drivers. It proved to be the Cobra's first FIA GT victory, Gurney winning outright. Holbert's engine did not last the distance, but Miles brought his roadster home to second place in the GT class.

The Cobras did not have the whole year their way, however, and at Nassau Chevrolet took its revenge. Three Grand Sport Sting Rays were given a thorough going over by the factory and ended up with engines developing around 500bhp in lightweight spaceframe chassis. Entered under the name of John Mecom, an enormously wealthy Texas oil man, the Grand Sports proceeded to pound the Cobras into the dust and disintegrating tarmac of the circuit. Although this was a bitter defeat at the close of the season, in real terms it was not very significant because the events in the Bahamas counted towards no championship and the Shelby successes of the year far outweighed this reversal.

Also running at Nassau were the cars that became known as the King Cobras. During the summer of 1963 Shelby had arranged with the Cooper Car Company of England to purchase a few of its rear-engined sports racing cars, which were normally equipped with a 2½-litre Coventry Climax engine. Shelby's intention was to fit the same Ford 289 V8 as used in the Cobras, and this was done in conjunction with a Colotti transaxle and with the tubular chassis suitably strengthened. What had been a fast car in 2½-litre form became blindingly so. And

ESSENTIAL AC COBRA

The great Ken Miles – wily racer, clever engineer, expatriate Englishman. After his off-course excursion during practice for the 1964 Sebring 12 Hours, he burns midnight oil with a few faithful mechanics to repair his leaf-sprung 427 Cobra so that it can compete in the race. At Watkins Glen the year before photographer Charles Justus Wick paid homage to Miles in his pencilled caption on the back of the original print: '66 laps, 2nd overall; 82 laps, 3rd overall; all in the same afternoon; what an iron man'.

in the car's first important race, the *Los Angeles Times* Grand Prix at Riverside, Dave MacDonald walked away from the opposition to win by a complete lap. This was a lucky victory, in fact, owing to the faster Chaparral of Jim Hall sidelining itself by self-incinerating its electrics while in the lead. Although the King Cobras were campaigned by the Shelby organisation they are not really part of the Cobra story, but these machines, and others like them, were the forerunners of the wondrous CanAm sports racing cars that rivalled Grand Prix racing for sheer spectacle and excitement through the second half of the 1960s and the early 1970s.

US events in 1964 would be attacked in the same manner as in the previous season except that the SCCA Class A production races had been altered to feature a series of regional championship races ending up in November with all the front-runners competing in a final at Riverside. The Cobras used for this season were, for the most part, built to exactly the same specification as the FIA roadsters. They featured enlarged wing flares with scalloped trailing edges to the doors to allow for them – cut-back doors in Cobra parlance. Each of these cars also had a bonnet air scoop, large vented brakes and peg-drive magnesium alloy wheels. The engines, with their quadruple twin-choke Weber carburettors, produced towards 400bhp and made these cars, along with the

RACING IN THE USA

FOLLOW THE EXPERTS

Castrol was the "oil of choice" used by the Shelby American Ford Cobras that won the 1964 road racing SCCA Manufacturers' Championship. This high-performance oil for today's high-performance engines belongs in **your** car. Follow the experts and protect your investment. Always ask for Castrol Motor Oil by name.

THE MASTERPIECE IN OILS
CASTROL OILS, INC., NEWARK N.J. 07105
SAN FRANCISCO, CALIF. 94109

other modifications, the ultimate Cobras in terms of handling, speed and reliability at racing speeds.

The Shelby team to contest both the SCCA United States Road Racing Championship and Class A Production Championship for 1964 was loosely made up of Ken Miles, Bob Johnson, Ed Leslie and Ronnie Bucknum, backed up by privateers such as Chuck Parsons and Lew Spencer. Bob Holbert and Dave MacDonald might have been with them, or could have gone to Europe with the coupés (see *The World Title Campaign*) after their good showing at Daytona and Sebring, but Holbert had a bad accident at Kent and poor MacDonald died on the second lap of the Indianapolis 500 when his car crashed in a ball of flame.

In spite of being overshadowed by the Shelby team's efforts in the GT World Championship, the roadsters carried all before them in US national events and Ed Leslie ended up SCCA Class A Champion. His first win of the season came at Tucson, Arizona, and he finished the year by winning the SCCA Race of Champions at Riverside to clinch the title.

During the 1965 season the new 427 Cobra was allowed to run as a Class A Production car in SCCA events, even though it had not been homologated by the FIA, and as a result the 289 was relegated to Class B Production. At the same time the legislating body disallowed the 427 from running with much of the tuning equipment, such as Weber carburettors, found on the 289. Thus FIA-specification 289s were capable of beating larger-capacity 427s on the track, although in these SCCA events they ran in a different class.

Cobras were once again all-conquering, however, in the SCCA Production and USRRC series. The factory continued to campaign 289s except for one team 427, which, most often driven by Ken Miles or Phil Hill, had to run in the modified class of the USRRC races and proved uncompetitive against the increasing numbers of mid-engined Group 7 cars.

The 1965 season proved to be the last for the USRRC series, as it was replaced for 1966 by the Trans American Sedan Challenge. But such was the success enjoyed by the now fully-sorted Cobras that they secured the title again, for the third year in succession, halfway through the season.

In spite of only that single 427 ever running as a Shelby team car, several privateers did use big Cobras. They swept the board in the SCCA Class A Production Championships from 1965 to 1968, and then again in 1973, nearly six years after the last one had left the AC factory in Thames Ditton.

RACING IN EUROPE

Quite apart from Cobra appearances in England and mainland Europe during the quest for the GT World Championship (see *The World Title Campaign*), the cars were raced extensively by private individuals and racing teams.

Before manufacture of the European series of leaf-sprung cars began, two cars from the US series, chassis numbers CS 2130 and CS 2131, became the first right-hand drive examples, specially made for racing and used in Europe. The first was invoiced to Peter Jopp and registered 645 CGT, and later became one of the John Willment racing Cobras. The second, registered 39 PH, was to be one of the Cobras raced at the 1963 Le Mans, this one run by a team sponsored by *The Sunday Times* and managed by Stirling Moss, who was still recovering

The Willment Cobra, chassis number CS 2131, muscling on at the Goodwood TT in August 1964. Jack Sears, truly one of the stars among Cobra drivers, is at the wheel. Seen the year before at Le Mans (facing page) is the Hugas/Jopp Cobra, complete with aerodynamic hard-top, powering through Tertre Rouge prior to building up to its maximum speed of over 160mph on the Mulsanne straight.

from the accident that put him out of Grand Prix racing. Another Cobra, chassis number CSX 2142, was to go to Le Mans that year, to be raced by Ed Hugas and Jopp.

Both Le Mans cars were fitted with racing 289 engines sent over from America. They also had bonnet scoops, perforated Dunlop competition wheels similar to

ESSENTIAL AC COBRA

The car sponsored by *The Sunday Times* at Le Mans in 1963, looked over at a press preview and seen on its way to seventh place for Bolton/Sanderson in the race. The cowl over the radiator intake, of the type used on racing Aces to improve top speed, was discarded for the race because it inhibited cooling.

Ed Hugas's Cobra during two of its outings in 1963. At Le Mans (right), where it is seen rounding Mulsanne corner with the fourth-placed Ferrari 250GTO, it retired with piston failure towards half-distance. At Snetterton (below) three months later Bob Olthoff ran it in the *Autosport* three-hour race, still in 24-hour trim with roof and number lighting, but retired with handling and engine problems.

RACING IN EUROPE

Sports car start at the Aintree 200 meeting in April 1964: despite his tremendous getaway, Sears could not keep ahead of Bruce McLaren's Cooper (right) and finished third. On tighter circuits, agility often gave more modern rear-engined cars an advantage over the Cobras...

those fitted on D-type Jaguars, and a special hard-top with a sloping rear that necessitated a two-piece boot lid. Other special features were extra spats to the rear of the wheelarches and an oversize fuel filler with a long neck that extended through the roof on the offside.

The English-sponsored car, to be driven by Peter Bolton (who had piloted ACs at Le Mans in 1957 and '58) and Ninian Sanderson, was painted green. The other car, which was largely sponsored by Hugas, East Coast distributor for Cobras and racing driver of some repute in the US, was painted white with blue stripes. In order to give it some racing mileage beforehand, this latter car was tried out by Jopp at Snetterton in full Le Mans trim. It had a pair of identification lights on the roof above the windscreen, whereas the English-entered car had only one such light.

At Le Mans the Hugas/Jopp car had to retire from 13th place when a piston let go just before half distance. The Bolton/Sanderson car, however, fared better and finished seventh, behind six Ferraris. It averaged 108mph and proved capable of over 165mph down the Mulsanne straight. At this juncture certain people involved with Cobras inevitably wondered what a more specialised version might achieve…

The two right-hand drive Cobras, one of them the successful Le Mans car, were later raced by John Willment's JW Automotive. CS 2130 was often driven by Bob Olthoff and it was this car which caught fire during the early stages of the Nürburgring 1000kms in 1964.

Willment made a sortie to Africa for the Kyalami Nine Hours race early in November 1963, sending Olthoff over for this. David Piper won in his Ferrari 250GTO with Olthoff coming a valiant second after rolling his car when a tyre blew. He somehow got the damaged car going again, persuaded it to drag itself to the pits for the wheel to be changed, and then heroically nursed the smoking heap around to just short of the finish line. There he waited until the chequered flag was out before crawling across the line for second place.

The Cobra was up and running again in December for the sports car race which supported the Rand Grand Prix, again at Kyalami. Olthoff was in the Willment Galaxie in another event so this time the Cobra was driven by Paul Hawkins, who won fairly easily against scant local opposition.

One of the first races of 1964 was the Oulton Park Trophy for sports cars and Jack Sears in the Willment

35

Three weeks before Aintree, Sears (above) caused a sensation at the Goodwood Easter International by battling hard for the lead with Graham Hill's Ferrari GTO. He had to settle for second place in his Willment-run Cobra (formerly that sponsored by *The Sunday Times*), which after this race sprouted a roll-over bar. One of the non-works cars in the Nürburgring 1000kms (left) was driven by Olthoff/Hawkins to 47th place after catching fire early in the race; lack of bonnet scorching suggests this is a shot from practice.

roadster managed to beat allcomers except for Jim Clark in a nimble Lotus 19. Olthoff was out in the other Willment Cobra for the Goodwood Whitsun meeting where he came third behind Roy Salvadori in a Cooper-Maserati and Hugh Dibley in a Brabham-Climax.

Meanwhile the AC factory's own racing Cobra coupé, developed to contest the 1964 Le Mans, had been taking shape. Rather than being taken out of the normal run, the chassis for this car was specially built and given the chassis number A 98. The bodywork, which had been designed by Alan Turner, was a beautiful piece of work and the whole thing looked like an expensive GT car that could in no way be mistaken for a Shelby Daytona coupé.

After much burning of midnight oil the car was finally ready for the road and was registered BPH 4B on 12 April 1964. Almost immediately it was taken over to France, as yet unpainted, for the Le Mans test sessions. On the Mulsanne straight it was timed at 249kph (154.7mph), which was only 11th fastest and slower than the lightweight E-type Jaguars, partially due to trouble with lifting at speed. But there was more to come…

A demonstration of this was given in the early hours of the morning of 14 June when the factory took the car out on the M1 and achieved some 185mph. As a result of

The rather lovely AC factory coupé, already with a broken headlamp fairing – minor damage considering the high-speed accident that was to come. In its wake in the colour shot is the Bondurant/Gurney Daytona coupé which was to win the GT class and finish fourth overall. Considering that a tyre failure caused the factory coupé to crash, resulting in the death of three spectators, Goodyear's 'success advert' – clearly illustrating the AC coupé – was the most amazing *faux pas*.

this exploit the company received all sorts of publicity, unintentional it said, and drew the Ministry of Transport's attention to the very high speeds possible by certain cars on Britain's new motorways. Within just three years the 70mph limit was in force.

By the time the car was taken to Le Mans again for the race, less than two months later, it was painted the traditional light green favoured by AC as well as Aston Martin for their racing cars. It has been said that the engine supplied to AC was down on power compared with Shelby's cars, having perhaps 30bhp less than the 385bhp or so that the American-entered cars were

37

More to placate the authorities at the Automobile Club de l'Ouest than for any other reason, Shelby entered an almost standard 289 roadster to be driven at Le Mans in 1964 by Frenchmen de Mortemart/Magne. They made no mistakes and drove a sensible race to bring the car home 19th.

rumoured to possess. Be that as it may, the British coupé ran with a low rev limit and higher axle ratio and achieved the same speed, 180mph, through the Mulsanne speed trap as one Daytona coupé, while the Amon/Neerpasch Daytona managed 186mph. Sears and Bolton were again paired for Le Mans and it was the former who put in the fastest practice lap at just over 126mph, only about 1mph slower than Gurney in his Daytona, which was the fastest GT car that year.

AC's coupé started the race with Sears at the wheel. Averaging over 120mph, he was mixing it with Gurney's Daytona through his opening stint, but shortly after Bolton took over early in the evening the car started to suffer from fuel starvation. After two pit stops to investigate, the fuel filter was found to contain some partially dissolved paper. Bolton then found himself able to maintain an average of just over 120mph in the dark until, on the 78th lap, the AC failed to reappear.

The unfortunate Bolton had lost control of the car when the offside rear tyre blew out between Arnage and White House. This occurred only a few laps after a fuel stop during which the Goodyear tyre people perhaps should have noticed that it needed replacing. Right behind the AC was a Ferrari 275P driven by Giancarlo Baghetti and in his efforts to avoid it he also careered off the road. Both cars ended up wrecked. The drivers were relatively unscathed but, by appalling coincidence, three Frenchmen had chosen to spectate by the side of the track in this prohibited area and the crashing Ferrari killed all of them.

So ended the factory's return to Le Mans. The remains of the car were returned to AC some time later, the French authorities having examined the wreck and tried to suggest that the limited slip rear axle had locked up. AC made no attempt to rebuild the coupé and never again raced a factory car.

One of Sears' most memorable drives in Willment's ex-1963 Le Mans Cobra came during the Ilford Trophy GT race, one of the supporting races for the 1964 British Grand Prix at Brands Hatch. Olthoff's practice time had made his car the fastest of the three Cobras entered for

Olthoff really gave his Cobra (car 27) a hard time during 1964, smashing it during practice for the Ilford Trophy at Brands Hatch, blowing the engine in Portugal, and modifying it thus during the Guards Trophy, again at Brands. Pristine by comparison is Lord Cross's Cobra (car 65), seen being thrown through a test during the Cambridge University Automobile Club's slalom meeting at Snetterton in late 1964.

Sears staged one of his famous spirited drives in the 1964 Goodwood TT. He came fourth despite spinning to last place on the opening lap – and picked up a few bashes.

this event, but he had badly damaged the car when he left the track and hit a tree. During the same session Sears lost a wheel on his car, but without too much damage. The third Cobra, a car that had been constructed for 'Tommy' Atkins, was so new that it was still unpainted, but Roy Salvadori, who often drove Atkins' cars, was only half a second or so a lap slower than the flying Olthoff.

As the cars took their place on the grid for the start of the race Sears found himself ushered to a less advantageous position than his practice time deserved, but there was no time to resolve matters before the start. Clearly displeased, Sears made the best of his situation and was very soon in second place behind Jackie Stewart in a Coombs E-type Jaguar, only to be shown the black flag at the end of the second lap merely for taking up, blamelessly, that incorrect position on the grid. Sears roared out to rejoin the race, now in seventh place, shaking his fist at the officials, while some of the Willment staff indulged in a punch-up in the pits out of frustration.

Sears then turned in one of the most notable drives of his career. Right on the limit and displaying single-minded brilliance, he started to reel in those in front of him. By lap seven he was second, and then on lap 15 he passed Stewart on the Portobello straight. He went on to win, to the ecstatic applause of all present. Salvadori was third in his first ever outing with the new Atkins Cobra.

Later in July Olthoff contested a GT race at Zolder in Belgium, his Willment Cobra repaired after its accident at Brands Hatch. For the first part of the race he hounded Lucien Bianchi in a Ferrari 250LM, and on the 21st lap eventually forced his way past on this rather twisty course. He proceeded to pull away but the Cobra's engine began to roughen up and he made a lengthy stop at the pits to try to repair the distributor. Out he went again after losing nine laps, but his race was run and he retired three laps later.

Olthoff was kept busy with a race in Portugal the following weekend but the engine failed, while a week later he and other Cobra drivers were at Brands Hatch for the Guards International Trophy. Both Sears and Olthoff used the former's car to qualify as Olthoff's was not yet ready after its Portuguese misfortune. The race was a close fight in the GT category between Sears and the young Chris Amon in the Atkins Cobra, with Amon making brave but unsuccessful attempts to pass the forceful Sears. They finished in fifth and sixth places overall, heading their class, with Tommy Hitchcock and Olthoff trailing back in 15th and 16th places. Hitchcock, an American privateer, managed a win in his left-hand drive FIA car at Mallory Park in a clubman's GT race during August.

By this time Cobras were beginning to find their way into lesser English club events. For instance, a pair of Cobras appeared at Croft in the hands of Keith Schellenberg and Bruce Ropner for a four-hour relay race during September, while Lord Cross was out in his newly acquired car for the five-hour relay race at Oulton Park in October.

The *Autosport* three-hour race at Snetterton saw another win for Sears, in appalling weather conditions which resulted in the race being terminated a little before time, on his first outing in the Daytona-like Cobra coupé that the Willment organisation had built. Amon came third in Atkins' Cobra roadster, behind Roger Mac's E-type Jaguar.

Rounding off the 1964 season, the Willment coupé was shipped to Africa for two race meetings held in November. The first weekend of the month saw it run in the *Rand Daily Mail* Nine Hours at Kyalami, where Olthoff/Sears finished fifth after a race not without incident. For much of the opening hour Olthoff sat in second place behind eventual winner David Piper's Ferrari 250LM, but then a rear tyre burst, the car spun off into the dirt, lurched onto its side and, mercifully, dropped onto its wheels again. After Willment's mechanics had changed the offending tyre out on the circuit, Olthoff drove the coupé back to the pits for all four wheels to be replaced – and for its leaking fuel tank to be sealed with the expedient use of some soap! – before resuming for a fairly uneventful run to the finish. Four weeks later at the

Night stop for the Willment coupé during the Rand Nine Hours in 1964. Driven by Sears (squatting on the pit counter) and Olthoff, the car finished fifth after a front wheel bearing broke following an earlier accident caused by a burst tyre. Willment Ford Galaxie driver Paul Hawkins is also seen, in the dark helmet.

James McNeille circuit, near Bulawayo, Olthoff entered two GT and Sports events supporting the Rhodesian Grand Prix, his coupé finishing second behind David Prophet's Lotus 30 in one but expiring in the other.

Probably the first races of any importance in England in 1965 were at Brands Hatch over Whitsun and included the Redex Trophy for GT cars. Roy Pike in the ex-Atkins roadster, which was now owned by Chequered Flag garages, had a ding-dong battle with John Miles in a little Diva-Ford but kept ahead to win. A week later at Silverstone it was the Eight Clubs meeting: Nick Granville-Smith and Lord Cross, driving almost standard road cars, both did well, the former coming third in a handicap and winning a scratch race with the car he had bought new in December 1964. At the mid-June BARC meeting at Silverstone Neil Dangerfield managed third place in a short race for big GT cars. Cobras made appearances at two more club events in July: at Mallory

RACING IN EUROPE

The Willment coupé had a lower and more purposeful appearance than the Daytonas from which it drew its inspiration. Both views show it during the 1965 TT at Oulton Park, where Frank Gardner finished fifth in the GT class and 10th overall. The editor of *Autosport*, Gregor Grant, must have been a fan of the car as he chose it for the weekly magazine's first ever colour editorial cover, even though this 25 June issue appeared nearly two months after the event it depicts...

Park Brian Wilson won the last race of the day, a handicap, while at Castle Combe Paddy McNally – an *Autosport* correspondent who came into the public eye two decades later for his relationship with the lady who became the Duchess of York – came second to Ron Fry's Ferrari 250LM in the Allcomer's race.

Later in July Silverstone once again hosted the Martini International Trophy for sports cars, run over 150 miles. At the drop of the flag Sears was left on the grid in the Willment Cobra with the throttle stuck wide open, which would almost certainly have blown the engine had he tried to start it. The trouble took a while to rectify and he was many laps in arrears before he got going. Although he then equalled the lap record for the class, he was too far behind to feature in the results. Pike, driving the Chequered Flag Cobra, upheld the marque's honours, however, with sixth place overall and first in the GT class.

Granville-Smith was also driving at the Martini

41

meeting and gained a second place in another race for sports and GT cars. The annual BARC meeting at Crystal Palace was hardly an auspicious occasion for Cobras, for Pike put a rod through the side of the block in the Chequered Flag car and Dangerfield collided with a TVR Griffith.

The Brands Hatch Guards Trophy meeting during August included the 20-lap Redex GT race. Bob Bondurant led from the start in the Chequered Flag Cobra followed closely by Sears in the Willment coupé, but by lap seven Bondurant's clutch was on the way out and the coupé assumed the lead. The race ended with Sears an easy winner and McNally 10th in a roadster, Bondurant having fallen right out of contention.

The organisers of the Australian Tourist Trophy, held in mid-November, wanted an attraction for the general public and persuaded Shelby to enter Ken Miles in the works 427 racing roadster. Miles was lying a close third behind a pair of Lotus 23s on this unsuitable, tight circuit when the rear suspension partially collapsed, in turn causing a rear tyre to burst. From there Miles went on to the Macau Grand Prix where he was paid to drive a 289 roadster by an enthusiastic private owner.

The Angolan Grand Prix at the beginning of December had Jo Schlesser driving another 427 Cobra and the enthusiastic amateur Keith Schellenberg in his 289 roadster that had been shipped from the UK for this

Two more Oulton Park TT scenes from 1965. Roger Mac's car was the fastest non-works Cobra until it lost the nearside rear wheel, but Sir John Whitmore's Shelby-entered roadster, seen with Gardner's Willment coupé tucked in behind, ended up winning the GT category.

event. Schlesser led from the start, but then traded the lead between David Piper in his Ferrari 365 P2 and Denny Hulme in a Brabham BT8. On lap 34 Schlesser's 427 suffered from the same rear suspension failure that had befallen Miles in Australia, but Schellenberg made 10th place.

By 1966 the day of the big names racing Cobras was over and the cars appeared in less important events. At

RACING IN EUROPE

Attention required by Gardner's Willment coupé (right) in the 1965 TT meant that it had to start late from the pits for the second half of this two-part race. Chequered Flag Cobras (below) in the paddock before the 1966 Guards Trophy at Brands Hatch: Mike Beckwith's 289 is in the foreground while Chris Irwin's 427 is almost obscured.

Brands Hatch in August the BRSCC meeting included a 20-lapper for GT cars in which Eric Hause inherited the lead from a spinning Elan but then lost it himself and finally finished fourth. The 750MC relay race at Silverstone in the same month attracted a team of Cobras from the AC Owner's Club, the drivers including Granville-Smith, Litchfield, Lord Cross, Harris and Lawrence, but they were on scratch and when rain started halfway through they were unable to overcome their handicap.

At a Silverstone club meeting at the beginning of September Bob Burnard with a 427 Cobra beat Fry's Ferrari 250LM in the GT race, and later that month at the Brighton Speed Trials he broke the class record in the same car with a run in 22.74sec. Burnard was not quite so fortunate at a clubman's meeting at Snetterton towards the end of September for his Cobra started to blow oil out of the engine and finally expired in a cloud of smoke. Burnard repaired his car for the last race of the season at Silverstone during October and won the GT race from Fry in a GT40, while Lord Cross managed sixth place in another sports car race. Later that month at Snetterton's last meeting Willment entered Brian Muir in the coupé for the Special GT and Marque race, and although Muir did win he was hounded all the way by Willie Green in a little Ginetta G12.

As can be seen from this far from complete but representative resumé of the competition activities outside America, the Cobra became obsolete quite quickly owing to the rapid pace of development in the field of sports racing cars during the mid-1960s. Their superior power was not enough to keep them ahead of more nimble, lighter cars that often had a better power to weight ratio. Cobras continued to be used at club level, however, and then transferred to historic racing during the 1970s.

43

The World Title Campaign

The racing Cobra roadsters had proved perfectly adequate for the short circuits upon which many of the US championship races were contested, their acceleration compensating for the indifferent aerodynamics of the bodywork that limited top speed to some 160mph. However, with the prospect of a serious attempt at the GT World Manufacturers' Championship, the various rounds of which were held for the most part at longer, less sinuous circuits, a higher top speed would be essential, especially since the main protagonist, Ferrari, had replaced its successful 250SWB Berlinetta with the even faster 250GTO that could top 175mph.

The hard-tops fitted to Cobra Le Mans and Le Mans replicas gave them a slight advantage over the normal open roadsters, but something more drastic was called for if Shelby's cars were to have any chance of success.

The Bondurant/Schlesser Daytona coupé during the Rheims 12 Hours in 1965 on its way to winning the GT category – and securing the GT World Championship title for Shelby.

Besides this, although a few Le Mans replicas were in the US, American drivers generally seemed not to get on with them and preferred to race Cobras in open form, even when pitched against the GTO, but they managed some decent placings in spite of the open car's inferior air penetration.

FIA regulations for Grand Touring cars stated that 100 examples had to be produced within a 12-month period, but as long as the bodywork conformed to general guidelines its precise form was the choice of the manufacturer. It was debatable whether the Ferrari GTO

THE WORLD TITLE CAMPAIGN

had correctly qualified under these rules, but it had already been homologated and that was *fait accompli*.

Carroll Shelby got together with two of his best men, Ken Miles and Pete Brock, and the upshot of their discussions was the initiation of the Daytona coupé project. The rolling chassis from the roadster crashed by Skip Hudson during the Daytona 500 was stripped of all extraneous tubework so that it could be used as the basis for a prototype. Among the modifications, Miles devised a lower seating position in conjunction with greater rake for the steering, while Brock began working on the shape of the bodywork. When his drawings – actually little more than sketches – were done, Brock began to build a full-size, three-dimensional template on the chassis with the help of two other Shelby employees, John Ohlsen and Bill Eaton.

The whole prototype 'kit' was taken to California Metal Shaping, a company that had also been responsible for the bodywork on the Scarab racing cars built by the former owners of Shelby's premises. Unlike the Italians, who preferred to make aluminium bodywork by beating panels around pieces of timber, this firm tended to produce sizeable panels on a wheeling machine and then weld them together, and it was by this method that the first coupé body was produced.

At the same time as this was in hand, Shelby had been in touch with the AC factory and asked for two cars to be flown over in chassis form. These were CSX 2286 and CSX 2287 and they left the factory on 22 November 1963. Upon arrival in California, the second of them was prepared to accept the coupé bodywork by the addition of specialised tubework, and in addition the radiator was re-mounted with a pronounced forward slope. When the body template was returned a multiplicity of 1in tubes were formed to support the aluminium shell and this chassis then had the panelwork erected upon it.

One of Remington's 385bhp engines was installed and mated to a Borg Warner T10 gearbox fitted with an

Details from a restored Daytona coupé. Carburettors are fully valanced in order to draw cold air, and the bonnet catches, rather amusingly, were borrowed from the humble Triumph Herald and carry 'M' for Michelotti. Stewart-Warner instruments were used, as on regular 289 Cobras sold in the US – but this is obviously a functional racing car.

45

ESSENTIAL AC COBRA

Daytona coupés under construction in the Shelby works during 1964, and the disastrous end to the first car's race debut. Driven by Holbert/MacDonald, the coupé was leading the 1964 Daytona Continental by five laps when its race was terminated by this fire.

'M' (or Sebring) gear cluster in conjunction with a rear axle ratio of 2.72:1. If it worked, this transmission combination would provide a theoretical maximum of more than 180mph.

Time was short before the 1964 Daytona Continental at which Shelby was hoping the car would make its first competition appearance, but a brief test was made at Riverside during February. The coupé handled well, attained over 160mph and gave markedly improved fuel economy, although cockpit heat was rather intense – but it all seemed promising. The car returned to the Shelby workshops for a coat of Viking Blue paint and final finishing before it was taken to Daytona.

The 1964 season

From the outset of practice at Daytona the new coupé proved itself to have the legs of the fastest of the Ferrari GTOs, let alone other entrants. But the slight concern regarding the cockpit heat that had built up during the short test at Riverside became a serious problem in the humidity of Daytona. The only stop-gap measures that could be adopted were quickly to cut some extra cooling apertures, after which the drivers – Bob Holbert and Dave MacDonald – decided they could just about bear it.

With the race three hours old the coupé was comfortably in the lead, and then the cushion was further extended when the quickest Ferrari lost time after bursting a tyre. By 600kms into the 2000kms race the coupé had pulled out a tremendous five-lap lead. Spirits must have been high in the Shelby camp as the car burbled into the pits for a driver change and fuel, but it became apparent that all was not well when Holbert explained that the rear axle was overheating. At the same time as the large fuel tank was being replenished, John Ohlsen crawled under the car to see if he could identify the problem. Then came disaster.

Some spilt fuel under the car ignited and the whole rear end, including the unfortunate Ohlsen, was suddenly engulfed in flames. Quick-acting marshals pulled him out and doused the flames, Ohlsen escaping with some painful burns. Burned-out hoses and wiring meant that the car was unable to continue.

The cause of the problem that Ohlsen had been investigating was traced to a faulty axle cooling pump that had blown an oil seal, so the car might not have lasted the

THE WORLD TITLE CAMPAIGN

The Cobra threat gradually disappeared in the 1964 Targa Florio after an impressive start, with only one of four roadsters entered reaching the finish. Arena/Coco (right) retired from fifth place with a broken oil pipe, while Masten Gregory (below) crashed the car he was sharing with Innes Ireland. Imagine racing past the bystanders in that village street...

distance anyway. Ferraris finished in the first three places trailed by Dan Gurney and Bob Johnson in a Cobra roadster, but such was the coupé's performance while it was running that from now on the aerodynamic closed versions were always known as Daytona coupés.

Before the next race, the 12 Hours at Sebring five weeks later, further attention was given to the problem of excessive cockpit heat with the addition of vents on the cowl. For this event the singleton coupé was backed up by four works roadsters and a special project car – a leaf-sprung roadster with big-block 427 motor – modified by Ken Miles under the auspices of Ford.

The same drivers as before, Holbert and MacDonald, were in the coupé. Nothing went wrong this time, except for the roasting the drivers had to suffer, and it finished first in the GT class behind three prototype Ferraris. Roadsters placed 2-3-5-6-7-12 in class, and but for a last-minute accident the coupé might have been denied its class victory by the quickest roadster.

Gurney and Johnson were in an incredible third place overall when, just as the Cobra was passing the pits with Johnson at the wheel, it came upon a partially disabled Alfa Romeo with only one feeble tail light showing through the darkness, crawling round to try to make the finish. There was no chance to take avoiding action and the Cobra, which was probably travelling 80mph faster than the Alfa, smashed straight into the back of it and somersaulted over the top. The Alfa spread petrol all over the place and the whole lot went up in flames, but the driver was dragged out relatively unharmed and the lucky Johnson was able to walk away from his written-off Cobra. Such had been their pace prior to this accident, however, that Gurney and Johnson still secured seventh place in the GT class and 10th overall.

Ken Miles did not have a happy time in the 427-powered Cobra. In practice he managed to overcook it and careered into one of the few trees around the course. The car was repaired overnight and managed to make the start, but apart from being the most tremendous handful it began to fall to bits. Finally the big side-oiler engine put a rod through the block, allowing a relieved Miles to call it a day.

The FIA events now moved to Europe with the Targa Florio, the notoriously arduous race held in Sicily over the mountainous Madonie circuit with a lap distance of 44.7 miles. This was not really a suitable environment for the Cobras, but the race was part of the World Manufacturers' Championship and some sort of attempt had to be made. Four roadsters were entered and at first did rather well.

After six laps Gurney and Jerry Grant were second overall and first in class, with Phil Hill and Bob Bondurant fourth and local drivers Arena and Coco fifth – but the fourth car, driven by Innes Ireland and Masten

47

ESSENTIAL AC COBRA

Gregory, had already crashed out. Further misfortune then struck as the locally-driven Cobra suffered a fractured oil pipe and expired, while the Hill/Bondurant car's suspension gave in. The Gurney/Grant car limped home, also with deranged suspension, in eighth place.

Spa-Francorchamps in May saw the coupé's first European appearance, in company with three roadsters. This very fast 8-mile track promised to be tailor-made for the coupé, and so it proved in practice with consistent lapping at over 125mph. But the race itself turned out to be a disaster with the coupé sidelined in the pits for five

The one Cobra roadster to survive the Targa Florio in 1964: Dan Gurney shared this car with Jerry Grant (driving here) to salvage eighth place despite suspension damage. After this experience, Carroll Shelby decided not to bother entering the 1965 Targa...

laps while dirt was cleaned out of the carburettors and fuel pipes. When he finally got going Hill pulverised the GT lap record with a speed of a fraction over 129mph, but it was to no avail and just two roadsters featured in the results, with Bondurant and Jochen Neerpasch sixth

THE WORLD TITLE CAMPAIGN

The Nürburgring was another circuit deemed not to suit the Daytona coupé, and another where the Cobra roadsters performed disastrously in 1964 – but at least the Bondurant/ Neerpasch car, pictured at the *Karussel* (below), set fastest GT lap before it retired from the 1000kms. Two coupés, however, were ready for Le Mans, and their drivers (right) are seen with the team boss: from left are Bob Bondurant, Chris Amon, Jochen Neerpasch, Dan Gurney and Carroll Shelby.

in the GT class and Jo Schlesser and Dickie Attwood eighth – a disappointing result.

The Nürburgring at the end of May was another unsuitable course for Cobras, or at least for the coupé. With Le Mans just a few days away, therefore, the closed car was sent to Ford France to be prepared for the 24-hour race. The tortuous course in the Eifel mountains took its first Cobra victim during practice when Arena crashed the car he was to share with Coco and badly injured himself. Although no-one else got hurt during the race itself, the whole outing was a disaster.

The Schlesser/Attwood car made a valiant effort to stay with Mike Parkes in a Ferrari GTO until its ignition wiring failed. This was repaired only for the carburettor linkage then to disintegrate, which meant more time lost in the pits. A fractured oil pipe finished the engine on Bondurant's car, while privateer Tommy Hitchcock's race ended when he was unexpectedly confronted by a gyrating Jaguar. John Willment's entry for Bob Olthoff in a right-hand drive roadster fared no better when it caught fire on the first lap. Although it was able to continue after fuel lines and some wiring were changed, it was right out of contention and finished five laps down.

The only car to finish was the Schlesser/Attwood entry in a pitiful 20th place in the GT class. The combination of reliability, speed and luck that had followed Shelby's racing Cobras during 1963 in US national events had well and truly deserted the team in its first full year of international competition. But then came Le Mans, and this was a special place for Shelby himself.

Soon after the first two chassis for this racing programme had been sent out from AC, two more – chassis numbers CSX 2299 and CSX 3000 – had been despatched, *sans* coachwork, to California at the end of November 1963. In time these two, along with CSX 2286, were sent to Italy to have Daytona coupé bodies built on them by Carrozzeria Gran Sport in Modena, and later the last two chassis supplied by AC to be equipped with coupé bodies – CSX 2601 and CSX 2602 – followed as well. These last two were outside the chassis number sequence of US leaf-sprung Cobras, the last of which was CSX 2589, but they left the factory – flown direct to Italy by British European Airways – in August 1964, three months before the final road car.

CSX 2299 was the first of these chassis to emerge from the Italian body builders and due to a misunderstanding over supplied dimensions the roof of this particular car was some 2in higher than on all other Daytona coupés. This was not the disaster it might have seemed as the slightly taller cockpit accommodated lanky Dan Gurney more comfortably, and so this unofficially became 'his car'.

49

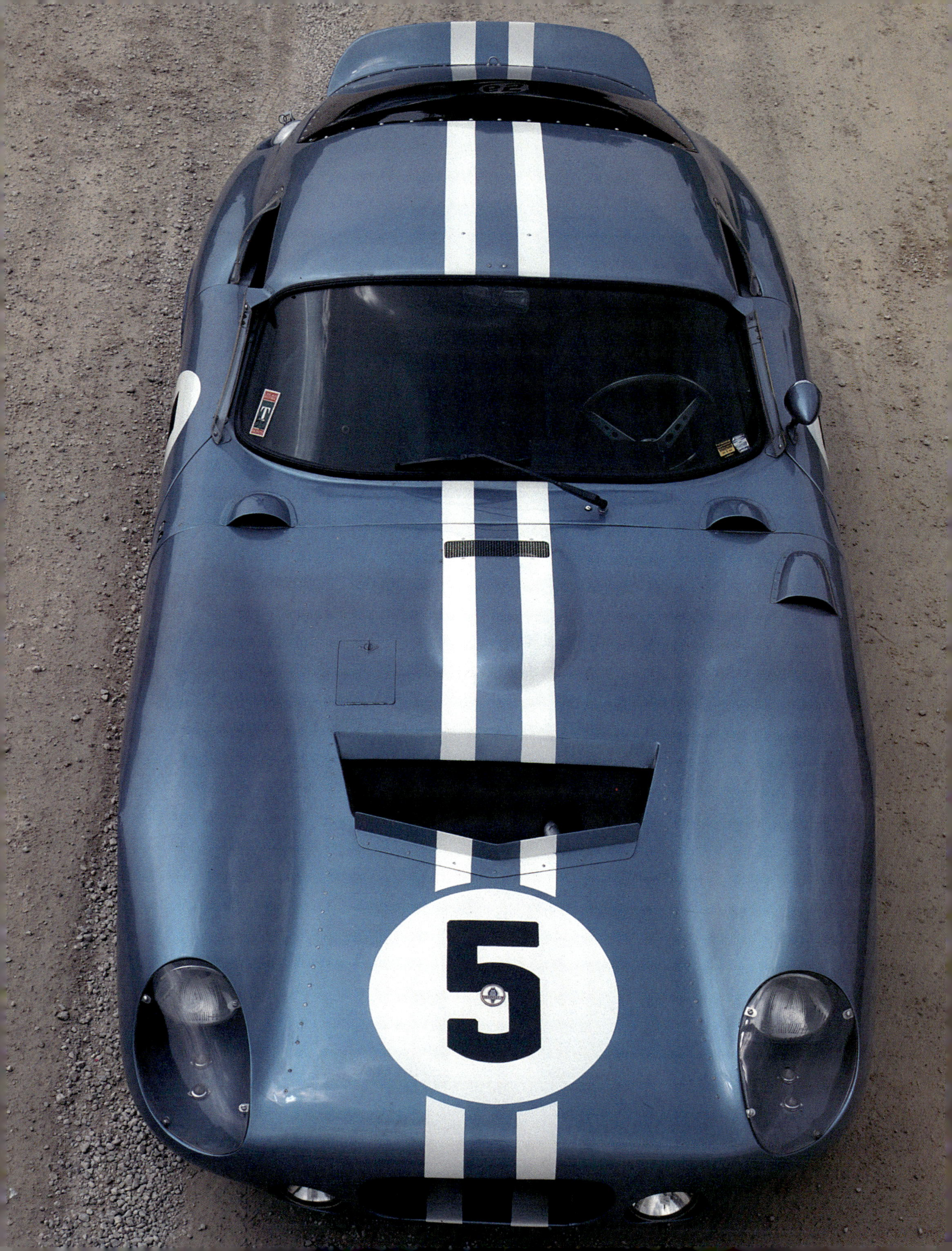

THE WORLD TITLE CAMPAIGN

The Daytona coupé comes good: three views of the Gurney/Bondurant car that won the GT class and finished fourth overall at Le Mans in 1964. High angle shows cooling vents for the cockpit, ducts in the side windows for the rear brakes, large radiator outlet duct and adjacent small hatch for checking engine oil without lifting the bonnet.

51

ESSENTIAL AC COBRA

Paddock workshop scenes of one of the Daytonas at Le Mans in 1964: relatively unsophisticated, but efficient, front-end layout with transverse leaf spring, and obligatory spare wheel carried at the rear.

Back to Le Mans. Two Daytona coupés were entered, CSX 2287 for Amon and Neerpasch and CSX 2299 for Gurney and Bondurant, and the AC factory's own interpretation of a Cobra coupé was also present (see pages 36–38). To the relief of the American contingent, their cars passed the sometimes fickle town-centre scrutineering that is a feature of Le Mans.

Also making their first appearance in the 24-hour race were the new Ford GT40s, conceived for the prototype class. Compared with the Daytona coupés, which began to look like creatures from another age, the three GT40s squatted low on the ground and gave new meaning to the word 'purposeful'. Apart from their V8 engines, they also bristled with innovative design features. It looked as though the Daytonas – contrived simply by putting an aerodynamic coupé body over an early 1950s sports racing car chassis that had lived on as the Cobra – might be made to look just a little foolish by the Ford, let alone by anything that Ferrari might pit against them in both the prototype and GT categories.

Both the GT40s and the Daytonas shared much of

THE WORLD TITLE CAMPAIGN

Two of the Le Mans class winner's pit stops: early on with the car still looking fairly smart (right), Gurney is in the car while Bondurant, far left, stands next to a pensive Phil Remington; more travel-stained at the final stop (below), Daniel Sexton Gurney to the fore.

their equipment and mechanic crews, not to mention the redoubtable Phil Remington, and they could rely on support from Ford Europe if needed. But even if it appeared that they were running within one organisation, the GT40s were already under the aegis of team manager John Wyer.

Practice times demonstrated that Shelby's behemoths had a little life left in them yet and in the first part of the race they showed the four Ferrari GTOs, which were in the same class, a clean pair of heels. So well did they go, booming down the long Mulsanne straight lap after lap, that as the night drew in they were running third and fourth overall and had a commanding lead in their class. The AC factory coupé had crashed in the dark after a tyre blew, but a pair of Frenchmen were still circulating steadily in a standard Cobra roadster supplied by Shelby.

At around midnight the leading Daytona, driven by Amon and Neerpasch, stopped at the pits for fuel and a driver change, but then a hitch occurred. The battery did not have enough power to start the engine. Quickly another battery was used to jump-start the car but in doing this the team broke the strict Le Mans rules – and so the car was disqualified. The surviving coupé of Gurney/Bondurant rumbled on through the night and then just as dawn was breaking the oil pressure gauge dropped to zero. Carefully the car was nursed back to the pits where a brief examination found that one of the oil cooler pipes had failed – the only cure was to bypass the oil cooler and risk a high oil temperature. The coupé ran on to the finish with a brief stop just before the end to try to cure a misfire caused by failing ignition points. Despite all this Gurney/Bondurant held on to fourth place overall and in the process won the GT class.

The Rheims 12 Hours was not a success. The same two Daytonas ran and both were eliminated by fractured gearbox tailshaft casings. Bondurant next entered the Freiburg hillclimb in Germany and then the Sierra

53

ESSENTIAL AC COBRA

Montana hillclimb in Switzerland. He won the GT class at both events, with Neerpasch and Jo Siffert, also in Cobras, not far behind him at Freiburg.

The next venue was the RAC Tourist Trophy at Goodwood in England. Gurney brought home the high-roof coupé first in the GT category, while Phil Hill in the older coupé had an oil pipe fail and so was only placed 11th overall. Sears and Olthoff made it a 1-2-3 for Cobras in the GT class with their Willment roadsters, but Salvadori in the Atkins Cobra retired with clutch trouble.

The Shelby team was now well placed to win the GT championship and a victory in any two of the remaining three events would clinch it. The Tour de France was next and, although the coupés did well in the early part, the Shelby back-up organisation simply could not cope with the enormous distances (the road sections amounted to over 6000kms) and the plethora of competitive stages (most of them over an hour long). Both coupés were out

Evocative Goodwood TT paddock view is the sort of shot most racing photographers never bothered with, but Geoff Goddard did. The nearest car is Sears' Willment roadster, looking rather archaic against the shapely Daytonas.

before the halfway mark and a third new coupé, CSX 2300, never even made the start because it was badly damaged on its transporter, en route from Italy, by the foolish expedient of driving under a low bridge.

The Coppa Europa at Monza was the most suitable remaining event for the Daytonas but, in a rather less than honest liaison between the organisers and Ferrari, it was cancelled. In short, the Italian manufacturer insisted that its 250LMs and 275LMs should be homologated as GT cars, and when this was refused the Monza organisers cancelled the race in support of Ferrari!

As an interesting aside to this, one of the other chassis

THE WORLD TITLE CAMPAIGN

Driven by Gurney, Daytona number 21 finished third overall and first in the GT class in the 1964 Tourist Trophy at Goodwood.

Into 1965, the second of two World Championship opening rounds in the US was the Sebring 12 Hours, where the heavens famously opened during the afternoon. Johnson/Payne finished second in the GT class and seventh overall.

(CSX 2601) in the process of being bodied at Carrozzeria Gran Sport in Modena had been slightly altered and fitted with a 427 engine. After the third coupé had been smashed on the way to the Tour de France, Shelby, who wanted all available cars to be at Monza, ordered that CSX 2601 be returned to standard 289 form.

Now with no chance of winning the championship, the Cobras still descended on Bridgehampton in September for the 'Double 500' meeting and took the first four GT places with roadsters driven by Ken Miles, Ronnie Bucknum, Bob Johnson and Chuck Parsons.

In their first full year of World Championship competition the Cobras had lost out to Ferrari. But not by that much. The score was 84.6 points to 78.3. It was at a press conference before the opening of his factory adjoining Los Angeles International Airport that Carroll Shelby is credited with the remark, "Next year Ferrari's ass is mine!" History does not record what Enzo Ferrari said about Shelby.

The 1965 season

For the 1965 season Shelby was given the task of running Ford's GT40 programme which, apart from the extra load on his workforce, meant increased finance for his company and also the chance to take on Ferrari in the prototype class as well as in the GT class with his Cobras.

From the outset there were problems with the FIA as both Shelby and Ferrari tried to homologate cars that did not comply with the regulation number that had to be built. Shelby for his part wished to get the new coil-sprung 427 Cobra recognised, but by the deadline only just over 50 cars had been built out of the obligatory 100. At least the AC company and Shelby had tried to meet the stipulation, but the FIA was becoming tired of Ferrari's antics and consequently was examining all applications with a more jaundiced eye. Previously Enzo Ferrari had done nothing other than present false paperwork to the effect that the necessary 100 250GTOs had been constructed whereas the

55

ESSENTIAL AC COBRA

Another of the four Daytona coupés entered for Sebring in 1965, this the Spencer/Adams car that finished fourth in the GT class and 21st overall.

reality was about 25. Then he had attempted the same ploy with the 250LM but had been refused. He tried again with this model and the FIA committee not surprisingly decided to scrutinise his affairs rather closely – he had lied again and the number built still fell far short of requirements.

The result of all this was that the two major protagonists went into battle in 1965 with the previous year's cars. Ferrari, showing his normal bad sportsmanship when thwarted, withdrew factory support for the GT series that year, leaving privateers and concessionaires to uphold the marque's name.

During the winter months all the Daytona coupés, which had been transported to the Shelby factory in California, were disassembled, modified where necessary and subjected to exhaustive testing, the latter with the help of the aeronautics department at Ford. The end result was a team of cars that might not have been at the forefront of development but were practically bullet-proof and as efficient as was humanly possible within the parameters of their design.

The new season opened on the last day of February with the Daytona Continental. Having the GT40s to look after as well, Shelby evolved a plan by which the Daytonas would pressure any Ferraris into being over-driven while the GT40s paced the other competitive runners to come through at the finish. In the event one Daytona blew its engine but the remaining three finished 1-2-4 in the GT class (2-4-6 overall), while the GT40s were first and third overall.

Sebring came next and the Shelby entourage fielded two GT40s and four Daytonas. That year's event was memorable, if for nothing else, as the day the heavens opened. At one stage during the race there was over a foot of water in the pit area, with general impedimenta and even wheels floating among the drenched pit crews. Cars still running were rapidly filling with water, open cars naturally being the worst affected but even coupés taking it in by the bucketful. Amazingly much of the field, including the four Daytonas, kept running through these appalling conditions, and when this trying race drew to a close the coupés were in first, fourth, sixth and 12th places in the GT class. One of the GT40s came second overall but the other fell by the wayside with collapsed suspension.

These first two US races of the new season proved that all the hard work over the winter months had been worthwhile. Improved ventilation provided greater driver comfort and therefore helped concentration. The speed, acceleration and reliability of the coupés were second to none, and the handling, although the transverse leaf springing gave an odd stance through corners, was sure-footed. To speed up pit stops, Shelby had devised and fitted integral air jacks operated by an air line that could be plugged into a socket on the nearside front wing. All of these aspects, plus the fact that the works Ferraris were out of the hunt this season, made Shelby's Daytonas an almost unstoppable force.

THE WORLD TITLE CAMPAIGN

Back to Europe again for the core of the 1965 season. Spa suited the Daytona coupés and they proved to be quick, but this car (right), driven by Sir John Whitmore, retired with a transmission problem. The Nürburgring 1000kms produced a stunning GT 1-2-3: victory went to Bob Bondurant's coupé (below), but he also did practice laps in the roadster (below right) that Shelby brought along as back-up.

The international GT series now moved to Europe, with the first race at Monza in Italy. Alan Mann Racing, run by one of the UK's leading Ford distributors and one that had a large competition department, were keen to have a part in things so two coupés and two GT40s were sent over to be prepared by them prior to this event. The Daytonas ran faultlessly throughout the race and came first and second in their class driven by Bondurant/Grant and Sears/Whitmore. The GT40s had a hard time due to Ferrari's return to the fray in the prototype class, but in any case the suspension on the Fords was still too fragile and one succumbed to this fault.

Oulton Park hosted the TT the following weekend. A Daytona was entered by Alan Mann for Jack Sears as well as the Willment coupé and four roadsters, two of the open cars entered by Radford Racing, one by Alan Mann and the other – the ex-Atkins car – by Chequered Flag. The race was run in two parts and at the start of the first Sears' car refused to fire, which meant a push-start from the pits and a two-lap penalty. Frank Gardner in the Willment coupé came out best in this with second place in the GT class to a Ferrari GTO, with Sir John Whitmore and Allen Grant behind him. Sears could manage no better than eighth in class, although he lowered the GT lap record four times during the race. The second part saw Sears lower the lap record yet again, but this time, without the penalty, he came third overall and first in class. The aggregate result for the two parts put Whitmore first in the GT class with Grant third and Sears fourth.

The beautiful, long, fast Spa-Francorchamps course once again denied the Daytonas a victory, although on paper it was eminently suitable for them. Bondurant finished second behind a privateer Ferrari GTO after spending time at the pits having a broken pushrod rectified, while the other coupé, driven by Whitmore, retired with a severe transmission vibration.

Shelby reckoned that the Targa Florio was not the place for Cobras – coupés or roadsters – after the previous

57

ESSENTIAL AC COBRA

The Nürburgring in 1965 looks alien to our eyes in these times of barriers and run-off areas – this is the second-in-class Sears/Gardner car. Three weeks earlier Sears had been entrusted with a Daytona coupé at the Oulton Park TT (left) where he broke the GT lap record repeatedly.

year's experiences, so this event was given a miss.

Three Daytonas were entered for the Nürburgring 1000kms at the end of May, two to be driven by Bondurant/Neerpasch and Gardner/Sears with the third, finished in Ford France livery, to be conducted by Jo Schlesser and André Simon. In practice Bondurant showed the car's pace by breaking the GT lap record and he followed this performance in the race by leading the other two cars home in a 1-2-3 GT clean sweep. This success was most certainly not shared by Shelby's other charges, the GT40s, with three retiring and one finishing in eighth place.

For the 1965 Le Mans 24 Hours Ford and Shelby put on a massive effort with five Daytona coupés and six GT40s, entered under various names owing to the limit on the number of entries from any one organisation. The

58

Despite a massive Le Mans effort with five Daytona coupés in 1965, Shelby could not repeat the previous year's class win. After halfway the cars wilted one by one, the Johnson/Payne entry retiring with a blown head gasket.

amount of money spent and the sheer number of cars should almost have guaranteed success, apart from one factor. The engine section of the Ford competition department had overstretched itself with its Indianapolis efforts and insufficient time had been allowed to prepare the new engines destined for the Le Mans cars. The Cobras at least were equipped with the proven 289 motor, but the GT40s variously used the 289, the untried 325 or – in the case of the two MkII cars – the 427.

By the time the race was at quarter distance all the GT40s had gone out but the Daytonas were looking very good – until the halfway mark approached. The Allen Grant/Schlesser car was first into the dead car park with clutch failure, then the car driven by English duo Peter Harper and Peter Sutcliffe. By 4.00am the coupé of Bob Johnson and Tom Payne had blown its head gasket, and next to go was the Gurney/Jerry Grant car also with a burned out clutch – which left just the Jack Sears and Dick Thompson coupé circulating. Its speed was severely restricted by some accident damage, sustained when it had hit a slower car, interfering with its cooling, but it managed to finish eighth overall and second in the GT class. Considering the number of cars running under the aegis of Shelby that year, this was one of his less impressive results...

In spite of the poor showing at Le Mans, victory in the GT championship was virtually in the bag for the Cobra Daytona coupés and they secured the title in the very next race, at Rheims. Just two cars were entered, still prepared by Alan Mann, for Bondurant/Schlesser and Sears/Whitmore. They finished in that order, first and second in the GT class.

There was no further need for the Daytona coupés to race as a team and, with the exception of the Coppa di Enna race in Sicily during August, they did not. Why Shelby entered this event is a mystery, for he had no need to and the course, not unlike the Targa Florio, was unlikely to suit the Cobras. However, when nothing other than honour was at stake, the cars perversely lasted the distance and came home first and second in class.

ESSENTIAL AC COBRA

Despite looking a bit of a mess in the dawn light at Le Mans in 1965, car 11 of Sears/Thompson was the highest placed Cobra in eighth place overall. Schlesser/Grant in car 12 were the first to retire with clutch failure.

Back in the US the season's final FIA race, which counted towards the GT championship, was the Bridgehampton 500kms. Patriotic racegoers might have liked the chance to see once again the now victorious Daytona coupés in full cry, but Shelby only wheeled out a couple of FIA roadsters. In the race Payne retired but Johnson finished seventh overall and first in the GT class.

For the racing Cobras, this was just the final flourish at the bottom of the page as far as international championships were concerned.

After a shaky start with its GT40 programme, Ford was absolutely determined to come back in 1966 and steamroller the opposition. Shelby was the man to do it, and all his time and the company's resources would in future be directed towards Ford's own sports racing cars.

60

THE WORLD TITLE CAMPAIGN

Le Mans disappointment notwithstanding, the Shelby team clinched the GT World Championship at Rheims, victorious Bondurant/Schlesser (car 26) and second-placed Sears/Whitmore (car 27) taking the flag together – but they were 30 laps apart! Goodyear this time got it right with the choice of picture for its celebratory advert...

Whatever has been said about Cobras over the years, ranging from uninformed hero worship to pseudo-intellectual scorn, the facts are undeniable. Within three years of its inception, the Anglo-American hybrid, whose origins were in the little AC Ace born in the early 1950s and an ordinary Ford pushrod overhead valve V8 engine, took on and humbled all other GT cars, including Italy's finest. The formula of a relatively large engine in a simple, lightweight car had been tried before, but never with such devastating success. It was in the nick of time: success for such a venture would have proved impossible a year or two later, due to the quickening march of time that resulted in ever more specialised sports racing cars such as those built for the forthcoming CanAm Challenge.

61

COIL-SPRUNG COBRAS

Genuine 427 Semi-Competition (SC) versions are like the proverbial hen's teeth, but one such car is CSX 3042. Exhaust and roll-over bar are correctly black rather than chrome, as beloved by so many imitators.

Although the rehashed Ace chassis with its transverse-leaf independent suspension had served well for the first Cobras, Carroll Shelby was keen to update the chassis and at the same time start to use Ford's 427 (7-litre) V8 engine to produce a fresh model. Also behind this train of thought was the necessity to build a required number of cars before a model could be homologated for racing in production classes.

The basis of the frame remained a pair of tubes, now set further apart and of 4in diameter as opposed to the 3in used on Aces and the previous Cobras. Coil-spring independent suspension was to replace the leaf springs used until now, and the work of designing the new suspension set-up was carried out at Ford under the watchful eye of the company's expert in this field, Claus Arning, who also used the most modern computer technology available at that time to assist in this task.

Gone were the fabricated sheet steel suspension towers at front and rear that had served as mounting points for the leaf springs, to be replaced by pairs of tubular uprights rising from the main chassis tubes and carrying the coil springs and upper wishbones. Bottom wishbone mountings were welded directly to the main chassis tubes. The front uprights were joined by trepanned sheet steel and diagonally braced with 2in tubes, whereas those at the rear were horizontally braced. Bracketry to attach both wishbones and coil springs was fabricated around the top of these uprights. There were four tubular cross-members and the scuttle hoop was diagonally cross-braced to the main chassis tubes.

The whole structure of the revised Cobra, therefore, was more substantial. In part this was necessary due to the

COIL-SPRUNG COBRAS

Although this car started life as a normal 427 sent out to Shelby in July 1965, it was later brought back to England by racing driver John Woolfe. No 427s were produced in right-hand drive form, so the dashboard Woolfe obtained from AC for his conversion was the distinctive AC 289 design – the car probably received its European-type chassis plate at the same time. Paint colour has also changed from red to black, but otherwise this 427 remains very correct, with original leather trim and seat belts.

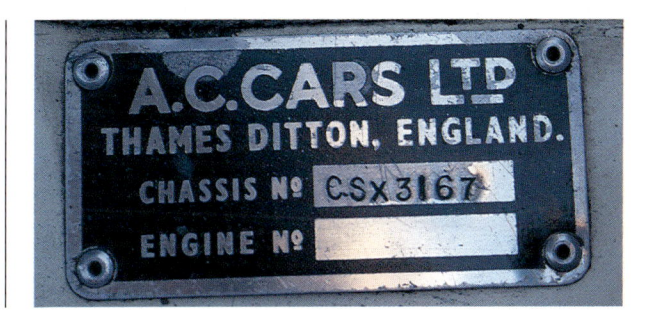

63

One of the few 427s produced with narrower rear wings between chassis numbers CSX 3125 and 3158. 'Sunburst' aluminium wheels are fitted. The 427 and Shelby's other wares were showcased (facing page) as the 'Cobra Caravan' toured major US cities – the signwriting on the truck says it all.

greater weight of the 427 engine, but more particularly because the greatly increased torque would be too much for the previous frame to cope with. This had been shown up a while earlier, at the end of 1963, when Ken Miles and a few others had installed one of these motors in a more or less standard 289 roadster still equipped with wire wheels. After some gentle testing, this car was modified with, among other things, increased cooling capacity, peg-drive alloy wheels and stronger leaf springs.

Brave Ken Miles elected to drive the thing in the 1964 Sebring 12 Hours. During practice he experienced its evil handling as he spun clean off the track into what was almost the only tree anywhere near the track. With the damage mostly to the bodywork, an all-night session put the car back into raceworthy condition – or, in the case of this particular Cobra, 'capable of being driven in a race' condition. Quite apart from the frightful handling, the car started to vibrate itself to pieces in the race, repairs in the pits becoming necessary as both brake and clutch mechanisms began to self-destruct. But Miles and co-driver John Morton were brave enough to extend the beast sufficiently to encourage the engine to put a rod through the side of the block, thus ending the debut of a 427 engine in a Cobra. This rather inauspicious experiment demonstrated that a new chassis was called for if the larger engine was to be used in a Cobra.

The reasoning behind the desire to fit this much larger motor was quite simple – Ford considered that the small-block 289 engine had reached the peak of its development. A 325 version, the absolute size limit allowed by this block, had been tried in the GT40s but had not proved a success, most likely because of insufficient development. Besides this, Chevrolet – the Cobra's main competitor in the US production racing series with the Corvette Sting Ray – could enlarge its small-block engine to 366cu in and was well into the development of its big-block 396-427 engine. Winning races sold cars and selling Cobras was of very little interest to the FoMoCo, but winning races with Cobras 'Powered by Ford' was of rather more interest because it helped sales of such ironmongery as Mustangs and Fairlanes.

The big-block Ford engine came in various forms, and the one developed for NASCAR stock racing, the 'side-oiler', was envisaged for the new Cobra. This motor had proved capable of producing a facile 480bhp and an equal figure of torque in pounds feet. Quite apart from Miles' experiences at Sebring, it obviously called for a radical redesign of the chassis frame.

As racing was to be the name of the game, it was necessary first to build the stipulated 100 vehicles to homologate the 427 Cobra as a *bona fide* production sports car, before any thought could turn to producing road versions. Accordingly, AC Cars Ltd produced the first two examples of the new cars and both were dispatched from the factory on 23 October 1964. The first one, in fact, was not a racing car and was shipped to Ford in Detroit for further testing and evaluation, but the second was designated as a 'race car' and went to Shelby in Los Angeles. They carried chassis numbers CSX 3001 and CSX 3002 respectively, and the latter was the only 427 that was specifically built for racing and campaigned by the Shelby organisation. It was equipped with a dry-sumped side-oiler motor and remote oil tank in the right-hand front wing, but it still had the smaller air intake of the 289 rather than the definitive gaping maw of the production 427.

It was not until 1 January 1965 that the flow of these homologation 427s from Thames Ditton began, chassis numbers CSX 3003 and CSX 3004 leaving the factory that day. By the end of the month cars up to CSX 3022 had been despatched to the US, and February saw another 25 cars completed. At that rate of progress the target of 100 was not going to be reached in time, but events that year, as described elsewhere (see pages 55-56), were to take a different course as far as homologation and GT racing were concerned.

During March another five cars left the factory before the attempt to meet the deadline was abandoned. The chassis numbers of these cars ran up to CSX 3055, with just one other chassis, CSX 3063, manufactured in this projected series of 100. This was made with a wheelbase 6in longer than normal and sent to Ghia in Italy to have a concept body made on it. In addition chassis numbers CSX 3027, CSX 3054 and CSX 3055 (the latter two both right-hand drive) were sent direct from the AC factory to Ford Advanced Vehicles in Slough and thence to Harold Radford, the coachbuilder. CSX 3054 had the Pete Brock-designed replacement for the Daytona coupé built on it, but the project was not completely finished at the time, while the other two chassis apparently never had bodywork made for them at Radfords. CSX 3055 was bought by John Willment, who oversaw the fitting upon it of a modified Ghia body from a 1950s Fiat 8V.

By the time the 427 competition roadsters began to arrive at Shelby's West Imperial Boulevard plant, near Los Angeles airport, he and others must have begun to wonder why he had bothered with such a project. A much more suitable machine for most SCCA events could be bought for just a bit more than the $10,000 or so that these Cobras cost, the homologation was about to fall through, and big-banger sports car racing was just entering a new era – customers in any significant numbers were just not interested in the new large Cobra as a sports racing car. However, a few did race, and very successfully, in the SCCA Class A production series, winning four years' running from 1965 to '68.

On 2 April the first roadgoing 427 roadster, chassis number CSX 3101, was despatched from the AC factory and flown by TWA to Los Angeles. It was sent to the US unpainted, but when cars began to leave Thames Ditton on a regular basis, some six weeks later, they were sent out already painted and almost invariably went by boat.

For years the roadgoing 427 roadsters, and the first run of homologation cars, have been talked about as 'MkIII' Cobras. This is totally incorrect. They were called MkIIs by the AC factory and this is clearly recorded in original factory documentation.

Of the earlier homologation MkII Cobras, just over a dozen found homes with people who wished to race them in SCCA events. The rest were simply left in the open outside Shelby's works, some of them even supported on blocks because their scarce Hallibrand magnesium alloy wheels had been used on other Cobras. While on a visit, an Eastern sales representative by the name of Charles Beidler noticed this quantity of cars just sitting there, without even a coat of paint on them, and voiced his opinion that they could be transformed into the ultimate road Cobras with a minimum of alterations.

Eager to see the back of his folly and with a profit beckoning to boot, Shelby rubbed his magic lamp, out popped a genie and with no more ado the revered 427 SC (Semi-Competition) was born. The genie, in truth, had little to do to effect this transformation. The side-oiler 427 engine was retained, but with normal iron cylinder heads and the compression ratio reduced to 10.4:1. A 'medium-riser' manifold with a pair of 4v 600 CFM Holley carburettors was standard, but many owners preferred the single 4v 750 CFM Holley as fitted to the actual competition cars. A thermostatically-controlled electric fan was fitted to enable the car to cope in traffic.

Whereas racing cars were fitted with bronze suspension bushes, the SC had these replaced by rubber ones, as used on the normal road cars. Hallibrand wheels – 7½in front, 9½in rear – were normally fitted, as on the later racing 427s, but shortage of supplies meant that some SCs had GT40-pattern wheels. Bodywork was left

ESSENTIAL AC COBRA

The Woolfe 427 retained a place in the *Guinness Book of Records* well into the 1980s as the world's fastest-accelerating production road car: 0-60mph in 4.2sec, 0-100mph in 10.3sec, standing-start quarter-mile in 12.4sec – no wonder passers-by stop and stare. Details show the following: location of stamped-in chassis number on coil-sprung cars, at right-hand front suspension mounting (CSX 3167 is the Woolfe car); the same number unusually stamped into the inlet manifold; a typical 427 chassis plate, with Shelby's exaggerated claim concerning manufacture; the type of filler cap used on 427s (and 289s), apart from racing and SC varieties; and the badge fitted to all 427s (and later leaf-sprung 289s).

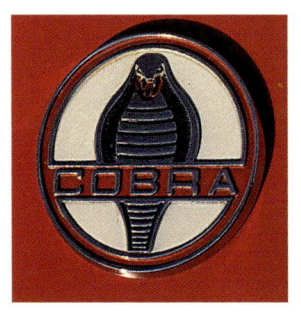

exactly as on the racing cars except for the addition of a small lip around the rear wheelarch and the standard fitting of a roadster windscreen, which was optional on the racing versions. Additional features to distinguish the SC from a normal 427 are the glass-fibre bonnet scoop with integral plenum chamber, external side exhausts, and visible riveting of the forward section of the bonnet to its frame.

Genuine SCs were only made using chassis numbers up to CSX 3055, with around 30 sold in total. No later cars with chassis numbers after CSX 3100 were ever supplied by the Shelby organisation as SCs.

The normal roadgoing 427 roadsters, which were now leaving the AC factory in quantities, had bodywork with the same configuration as the racing cars, including the large wheelarch flares, but lacked such fittings as the bonnet scoop and side exhausts. In May 1965, starting at

66

Two 427s received very distinctive bodies, both from the house of Ghia. For CSX 3055 John Willment fitted a modified body from a 1950s Fiat 8V. CSX 3063, with a wheelbase 6in longer than normal, was given this rather more successful – and thoroughly modern – shape as a styling exercise.

chassis number CSX 3125, narrower rear wings with less extreme flares were incorporated into the bodyshell, but after chassis number CSX 3158, which left the factory on 24 July 1965, the previous wide wings were once again used. The reason for these changes is that early cars were fitted with surplus racing bodyshells, and when these were exhausted the factory reduced the width of the rear wings to suit the narrower rear wheels of road versions. Due to customer demand, however, the wide wings were reintroduced, in spite of their unfortunate appearance if standard wheels are fitted.

The side-oiler 427 engine was not fitted to road cars as standard, the top-oiler version of Ford's big-block motor initially being used instead. Whereas top-oilers route lubricant to the cam followers and filter before the crank, side-oilers have the crank fed directly from the pump. An easy distinguishing feature between these two engines is that the core plugs of the side-oiler are of the screw-in variety while the top-oiler has press-in ones. Normally these engines were equipped with low-rise heads and inlet manifolding which allowed the fitting of a pair of Ford Autolite carburettors; these are not the best set-up as they tend to produce chronic fuel starvation on fast corners, so they have sometimes been replaced by alternative manifolding and carburettors over the years.

Both top-oiler and side-oiler motors share the same internal dimensions with a bore and stroke of 4.23in by 3.78in (107.4mm by 96.0mm). The crankshaft runs in five cross-bolted main bearings and is of cast iron, although from 1965 an alternative cross-drilled steel crankshaft, known as the 'Le Mans', was available and used in conjunction with different connecting rods. Overhead valve gear is normal rocker shaft type, operated by pushrods and solid tappets.

Neither of these 427 engines was exactly plentiful and supplies to Shelby, in spite of his relationship with Ford, were somewhat unreliable. It was probably this factor, but also maybe the temptation to make extra profit, that led Shelby to start to use the 428 'Police Special Interceptor' engine in many 427s. This motor was of the same family as the 427 but had a smaller bore and longer stroke at 4.13in by 3.98in (104.9mm by 101.1mm). The main bearings lacked cross-bolted caps and the valve gear had hydraulic tappets. Quoted at about 350bhp, power output was considerably down on the 427 motors, but the torque figure was similar.

It should be mentioned here that power figures quoted for American engines are best viewed with some cynicism. Were I a gambling man, I would place a sizeable bet that no standard 427 engine would ever develop the often-quoted 480bhp, or anywhere near it, on a Heenan & Froud dynamometer.

Although the 428 engines were invoiced to Shelby by the Ford Motor Company at around half the price of the 427s, and consequently he made more profit from cars equipped with them, various later MkII Cobras were fitted with 427 engines, seemingly at random but probably for favoured clients.

The gearbox used in 427s was an ordinary Ford top-loader, so-called because its construction meant that gears and shafts were assembled through the top aperture. With a cast iron case and stronger gears than in the Borg Warner 'box fitted to earlier Cobras, the top-loader was very heavy and gave quite a slow change, but it did prove man enough for its job. The three or four cars fitted with automatic transmission included a pair of supercharged cars which Shelby had built for himself and a customer.

Interior trim was much the same as for the earlier Cobras apart from the seats being slightly larger. Windscreen and weather equipment were also the same except for the back of the hood being tailored differently to allow for the rear wing shape of the 427 in both guises. Normal road cars were usually equipped with ten-spoke 'Sunburst' wheels of 7½in width front and rear, but some of the narrow-wing cars were sent out with GT40-pattern wheels of the same size.

During 1965 100 roadsters were shipped from Thames Ditton to Shelby's plant in California for finishing. This took production up to chassis number CSX 3201, which, in typical AC fashion, left the factory on 24 November whereas some earlier chassis numbers went out during December. Of that year's production, one car, chassis number CSX 3150, was kept back in England, initially consigned to Ford Advanced Vehicles at Slough and then exhibited in Belgium at the Brussels Motor Show.

A new contract with Shelby for another run of cars started at chassis number CSX 3202, which was also part of the first 10-car batch of 1966 cars to leave the factory on 10 January. Production continued in this way through 1966 with cars leaving in further batches for California by sea, a typical sample aboard *SS Risanger* during April consisting of two silver cars, four red, three green and one white, all with the black trim used on 427s.

The end was nigh however. Shelby and Ford had moved on and the Cobra, although it had served its purpose by winning the GT championship in 1965 and had garnered some sales for Ford, was now redundant and had to make way for Shelby/Ford Mustangs and the GT40s. The last batch of 427s, nine of them, left the factory on 28 December 1966 aboard the *Loch Loyal*. Two other chassis numbers, CSX 3359 and CSX 3360, are listed in the factory records but have no delivery date beside them, so they most likely were never completed at Thames Ditton.

Five 427 chassis from the 1966 production run remained in Europe. CSX 3217 went to Ford Advanced Vehicles where it was converted for competition use, and thereafter it was raced and hillclimbed in Europe. CSX 3222 and CSX 3301 also went to Ford Advanced Vehicles and thence to Europe, the latter specifically to France. Between CSX 3174 and CSX 3175 the factory slipped into the production run a couple of chassis with inconsistent serial numbers, CSX 5001 and CSX 5002, which both went to Ghia in Italy to have special convertible bodies built on them.

Although the 427, and more especially the 427 SC, is regarded by many as the ultimate Cobra and is the version most copied by the various – and for the most part fairly ghastly – replicas available around the world, there are many people, myself included, who prefer the earlier leaf-sprung version.

Somehow the handling is more responsive on the smaller-engined, lighter car. In spite of the more sophisticated chassis and suspension of the 427, the earlier car has more forgiving handling and you are less likely to get into trouble with it. Of course a standard 427 is a faster car, in sheer speed and acceleration, than an unmodified 289, but the smaller car, driven well, would have little difficulty in keeping up with its bigger brother on a normal give-and-take road. A tuned 289 almost certainly has more usable power than the 427, which, if pressed, has a combination of power and torque that is rather too much for its size and weight.

The looks of the 427 also are very much a 'love it or

Carroll Shelby's personal car, photographed during *Road & Track* magazine's 1968 road test of it. It has customised exterior features, twin Paxton superchargers and automatic transmission – the inspiration for many a replica?

leave it' affair. I myself find it an over-blown, somewhat ugly transformation of the early roadster. This is especially obvious when a car is fitted with the standard small wheels in conjunction with the wider rear wings that were used on the vast majority of 427s.

In 1974 *Road & Track* magazine road tested a second-hand 427 equipped with a side-oiler engine. The writer, who incidentally owned the car, commented on the offset pedal position, cockpit heat and rather primitive weather equipment. Steering and brakes came in for much praise, as did the gearbox. Actual performance figures were a compilation of earlier results and what was described as educated guesswork, the magazine admitting that it had never tested a standard 427 previously. A top speed of 162mph was quoted, with 126mph, 113mph and 77mph available in the indirect gears. Among the acceleration figures were 0-60mph in 5.3sec, 0-80mph in 9.4sec, 0-100mph in 13.0sec and the standing-start quarter-mile in 13.8sec – fast but no improvement on a 289.

In 1968 *Road & Track* had tested Shelby's personal 427 with twin Paxton superchargers and automatic transmission. Shelby claimed a power output of 800bhp! Be that as it may, this gruesome vehicle, complete with chrome side exhausts and roll-over hoop as well as an outsize bonnet scoop, managed 0-60mph in 3.8sec, 0-80mph in 5.6sec, 0-100mph in 7.9sec and the standing-start quarter-mile in 11.9sec. The top speed was given as 182mph but the writer was honest enough to admit that this was calculated on the gearing at 7000rpm.

It is a little difficult to arrive at a consistent set of figures for the 427, as can be seen from the test carried out on a 428-engined car by *Car and Driver* magazine in 1967. Remembering that this car had the less powerful 'Police Interceptor' engine, the results are most amazing: 0-60mph in 4.3sec, 0-80mph in 6.2sec, 0-100mph in 8.8sec and the standing-start quarter-mile in 12.2sec. There were again comments about cockpit heat and the observation that the car had 'a top that requires a degree in structural engineering to understand'.

Americans seemed generally unhappy about Ace and Cobra soft-tops but in practice the frame is simple to erect. The top fits simply and the parts stow away when not in use, giving a neat body line. When raised, the top is very evenly tensioned and the car can be driven at well over 100mph if you can stand the cacophony produced by a mixture of wind, mechanical and exhaust noise. It is even pretty water-tight apart from the odd drip – more water always comes up through the floor in a downpour.

ESSENTIAL AC COBRA

Top engine specification for a 427 is the side-oiler, developed for motor racing and, in Cobras, usually reserved for racing and SC versions. Dashboard view of the same car, in fact a roadgoing version, shows the Smiths instruments used in 427s.

The European AC 289

During the production run of the 427, which almost without exception was sold in the US, the AC factory began to make a small number of what it named the AC 289 Sports. Although the Cobra name, owned by Ford, was free for AC to use, the company decided not to. One might ponder that the hierarchy at AC, a little hurt or maybe displeased by Shelby's cowboy tactics to eradicate the AC name from what he considered his baby, decided to give this swansong Cobra their own name.

These final UK and European specification cars had chassis numbers with COB and COX prefixes, signifying right-hand and left-hand drive as before. The first car to

Somehow the wire-wheeled AC 289 – the 'post-427' model developed by AC Cars for the European market – looks more the archetypal 1960s sports car than its broad-shouldered, aluminium-wheeled counterpart. Rubber covering on leading edge of rear wing is owner-added protection against stone chips. Unlike the 427, the dashboard of this model featured a lower central section to provide room for a clock, as on earlier leaf-sprung cars.

be registered was the factory demonstrator, chassis number COB 6106, on 13 October 1965. Chassis numbers actually started at COB 6101, but this one did not leave the factory until 27 June 1966. The last complete car was COX 6125 but two more incomplete ones were produced: COX 6126 was sent to the US with no engine and COX 6127, which according to records was little more than a chassis and some assorted parts, went out on 15 July 1968.

The AC 289 was only produced in narrow rear wing form, as used on just over 30 of the first contract batch of 427 road cars. This gives these cars a more balanced overall appearance than the majority of the American versions. AC 289s were also always fitted with wire wheels, which to my mind suit the car, whose styling was definitely rooted in the 1950s, rather better than the various alloy wheels used on the 427s.

The combination rear lights used on all Aces (except the very early 1953 cars) and Cobras had now been replaced on later editions of the 427 and this model, apart from the prototype, by a separate rear/stop light and indicator on either side. Cobra badging did not feature on this car, the die-stamped AC badge used on the boot lid of the European leaf-sprung cars being employed on the nose as well.

The 289 engine used in these cars was the later variety with J-type cylinder block. The main difference from the earlier motor was that the gearbox was now attached to the bellhousing by six bolts instead of five. The gearbox continued to be the Borg Warner T10.

In 1967 *Motor* magazine road tested the prototype that had been used as factory demonstrator and was by then two years old. The writer opened his prose by describing it as 'the most splendid vintage sports car, a car that is characteristic of the image of the true masculine fun car'. Further extracts are worth quoting.

'The 289 stands on its own as a car whose tremendous performance and roadholding is forever a joy. Unfortunately the drawbacks are there too: it has a ride which many would consider poor; a primitive hood which you leave down unless it is actually raining when you are stationary; and little useful luggage space…

'With 271bhp in 21cwt of car, the 289 is the fastest car to 100mph that we have tested… The unit is extremely tractable and it has an excellent gearbox. Although the cornering attitude of the car is largely dictated by the throttle at road speeds, the way the power is put through to the road, particularly on wet surfaces, is quite remarkable. At high speeds, around the 100mph mark, it did not seem quite as stable as expected, but this was possibly due to worn wishbone rubbers on a two-year-old demonstrator…

'As lovers of high performance we soon became attached to the 289 particularly with the hood down to reduce engine noise and hood flap… There is some tappet noise, mainly audible when the hood is up, but otherwise the engine is fairly unobtrusive. In normal road use you won't use more than 4000rpm – if you wind on past 5000rpm towards the maximum of 7200rpm on this seemingly unburstable engine then it becomes more frenzied but never alarmingly so…

'During our acceleration tests we used up to 6500rpm reaching 60mph in just 5.6 seconds, 80mph in 9 seconds and 100mph in 13.7 seconds with a standing-start quarter-mile in 14.4 seconds. For our maximum speed we had to retire to foreign parts; the mean 134.9mph represented 6300rpm which was ear shattering with the hood up…

'Fuel consumption over that 640-mile trip worked out at 16.3mpg with a lot of cruising in the 80-100mph range… the gearbox has such pleasant ratios and the short stubby gear lever is so pleasant to slide around…

'The pedals are well placed for heel-and-toeing and there is room for the left foot beside the clutch, resting on a piece of floor which gets uncomfortably hot at times. The clutch itself grips smoothly and well but its long travel more or less dictates the seating position, and it is certainly the heaviest we have met for a very long time.

'It takes a little time to be able to use the 289 chassis and power to best advantage: you start by coming into corners too slowly and then blasting out too quickly. The tail is unlikely to unstick even so, but there are smoother ways of getting round. It pays to get the braking done beforehand, then go round the corner under more or less constant power until the exit is in sight. You can then use more throttle keeping the tail just in check, or let it come out a fraction lined up ready to squirt down the next straight. If the tail comes out it is easy to keep under control with a little less throttle and a bit of correction…

'On wet roads we learnt to be careful having at one point lost the tail while pulling out to overtake. It came back as soon as the power was eased but we developed the habit of using one gear higher than we would on dry roads; this reduced torque suitably without sacrificing too much of the phenomenal acceleration.

'This is all helped by the excellent steering which gives just the right kind of feel for safe fast driving. There is kickback on bumps – but it is nicely geared and surprisingly light at parking speeds. At very high speed very little movement is required and you are hardly conscious of doing more than vary the pressure on the steering wheel to get round quite a noticeable 100mph corner. This extreme sensitivity at high speeds around the 100mph mark leaves no room for heavy hands or feet; we were also a little perturbed at the tendency to veer to the right when power was eased after a full-throttle blast at high speed.

'This directional variation is due to rear wheel steering; we suspect that newer cars, or cars that haven't had a long spell as demonstrators, would not exhibit this feature, since it is almost certainly a sign of worn wishbone bushes.

'On the road the brakes feel heavy (no servo is used) but firm and fade free… At speed on the main roads it is

Motor's road test of an AC 289 in its 14 October 1967 issue was the best contemporary assessment of the Cobra. It looks like the tester got to grips with the car...

comfortable. On poorly filled-in roadmender's holes you get the odd rattle from the rear suspension, but otherwise the car feels rigid and free from scuttle shake on surfaces which might be expected to produce it.'

The test ends with the following rather amusing observation: 'The jack has to fit under the chassis tubes or under the wishbones, but with a flat tyre there is little room to get under: you need to spread the tonneau cover out on the ground, lie down and grovel underneath.'

I have quoted this road test at some length because it is one of the few magazine evaluations of the Cobra which I feel truly looks at the car objectively and genuinely gives an idea of the car's character. For instance, rear-end steering is a characteristic of Cobras, and Aces too for that matter, if the rear wishbone bushes are worn.

The temptation to launch into some kind of trendy 1960s journalese was too much for some scribes. The epitome of this must be one Jefferson Howard Jr when he wrote about the same factory demonstrator AC 289 (registration KPD 150C) and saw it thus in his piece entitled *Sweet Dreams*.

'Every once in a while a dream comes true. I mean, someone has to win the pools each week. Then there was this car, see, kinda mean, you know, with great bulges over its slicks and a big, wide mouth.

'I nearly missed it. They gave up production last year, but down on the Thames beyond Kingston there was one left in captivity. It has silver blue paint finish which nearly jumps out at you with glitter. A weekend at the wheel was like a date with Brigitte Bardot. I mean wild, man, as wild as you can get.

'Push that stump of a gear stick into bottom and tread on the loud. Hold tight 'cos you've no g-suit and this is Cape Kennedy.

'Ever heard about those three dots on your side of the solid white line? They're for Cobra power, aerosol urge, and this brute can do it.

'Up back you leave them in a kind of thunder. Like a swarm of ton-up bikes on a bender. Maybe the windows don't rattle but they oughta. It starts as a waffle, then in comes the crackle and from then on it's power waves, kapow! On a tight rein there's a tinkly threshing of machinery, poised. Supertaps open, bombs away and there she was – gone; with just the echo where she used to be.

'Can't be all good. You're right, cooky. You arrive everywhere hot-footed with those metal pedals glowing cherry.

'Comes on to rain, and there's that Chinese puzzle of a do-it-yourself roof kit.

'Drips on your knees, steam all round, and hot, hot, hot. But who cares? This is a go-car. It's for driving. Whenever you feel tensed and confined, just blow, man, up the road and back in the Cobra and you can't help but feel better.

'Unobtainable and unattainable, out of reach like the stars. I gave it back to the man downtown. It's his prize and he won't part with it again, unless he goes crazy or something. Anyway, it's worth three grand, and we don't have that kind of money.'

All I can say is that it is most fortunate that Mr Howard Jr, and hopefully others like him, could not afford a Cobra. It is nonsense like this, however, which helped to give the Cobra an undeserved reputation in some quarters and in all probability hastened the coming of the replica cult to satiate like-minded individuals.

Be that as it may, after chassis number COX 6025 no more complete Cobras left the AC factory at Thames Ditton, but after the two last chassis numbers (COX 6126 and COX 6127) five more chassis were built (numbered 6128-6132). These had a longer wheelbase and were

73

ESSENTIAL AC COBRA

Narrow rear wings make the swansong AC 289 look less bulbous than the majority of 427s. The AC factory proudly substituted Shelby's badge for its own identity on the nose, and twinned lights were employed for the tail.

supplied to the makers of the film *Monte Carlo or Bust* so that they could have vintage-type bodywork built on them, and as such they featured prominently in this film.

Subsequently some of these chassis have been shortened to Cobra length and had Cobra-type bodywork mounted thereon. Obviously any of these are preferable to an out-and-out fake, but whether or not they can be regarded as proper Cobras is debatable. The argument for them would be that the chassis were built by the factory at the end of the AC 289 production run, the case against that they were of a different chassis length and were never intended by the factory to be Cobras.

AFTER THE COBRA

'Imitation is the sincerest form of flattery'. But the imitation is hardly faithful in this case: steering wheel excepted, the tasteless interior of this Autokraft 'MkIV' belies its antecedents.

For a good many years after Cobras ceased to be produced at Thames Ditton, servicing and repairs could be carried out on customers' cars at the factory if required. One or two complete rebuilds of crashed cars were undertaken and spares were also in good supply, the latter in part due to the production of AC 428 Fruas, which shared many mechanical and chassis components with coil-sprung Cobras. However, when manufacture of these cars stopped in the early 1970s and the factory began to concentrate on its next project, which was to become the mid-engined ME 3000, the ready supply of parts for Cobras began to dry up.

Engines and gearboxes had always been obtainable in the US unless you wanted a genuine Hi-Po ('High Performance') 289 or side-oiler 427, which were elusive due to the limited quantities that had been built. Some of the gear sets for the Borg Warner T10 gearbox, such as the close ratio 'M' (Sebring) or 'K' with high bottom gear, were also becoming very difficult to obtain. The Shelby organisation had never carried a vast quantity of spares for Cobras and in any case Shelby's interests were by now elsewhere.

As a result of this various individuals in the US began to specialise in the repair, rebuild and remanufacture of parts for Cobras. Over the years, as the cars became older

75

and in need of more attention, and also as their value began to increase dramatically, a series of small industries grew up to serve owners and aspirant owners. As prices have inflated over the years, allied to the fact that there simply have not been enough cars to go round, there have grown up three Cobra-orientated offshoots.

The first I personally find rather tasteless but is in reality quite harmless – the plethora of Cobra replicas that have become available over the past 20 years or so. These range from rather poor glass-fibre lookalike bodies usually mounted on a proprietary floorpan from a mass-produced car, sometimes even retaining this car's small, low-powered engine, to full-blown attempts to recreate the chassis and running gear, at least in part, of the original, clothing it in a good quality glass-fibre body or even one of aluminium and fitting an American V8 engine.

Some pains are now taken with a few of the replicas to try to copy original instrumentation, and windscreens and other hardware have long been available. An indication of the scale of these operations is that one small business known to me has produced, in a period of a few years, around 2000 windscreen frames. Although these are occasionally bought by owners of real cars needing replacements, this number is approximately enough to equip every genuine Cobra ever made with two windscreens! And there are several other firms making these components both in Britain and abroad…

Some firms manufacturing these replicas, usually in kit form but sometimes fully built, have been in business for several years, manufactured a good number of cars and achieved an air of respectability, sometimes aided by endorsement from the likes of Carroll Shelby or John Tojeiro. They also have an enthusiastic following, with some products considered nearer the real thing or of better quality than others. But make no mistake. These cars have nothing whatsoever to do with AC Cobras and never will do. They have nothing more to do with the originals than a plastic replica Chinese figurine in a gift shop has to do with the priceless original in a museum.

The second Cobra offshoot is rather less savoury but has been perpetrated by a good number of individuals, in many cases several times and often quite blatantly. This is the out-and-out faking of cars. It has normally begun with the supposed finding of the remains of a Cobra. This Cobra can be found in several forms and it is truly a credit to these persons that they can locate cars in absolutely extraordinary circumstances and in almost unrecognisable or 99 per cent incomplete condition. Amazingly, by some lucky twist of fate, however atrocious the condition of the pitiful remains, there is always some way of identifying the pieces of metal as a particular chassis number! Cars also seem to become separated into major component parts and then turn up inexplicably in totally different locations. The other ploy is to acquire what Americans call 'title' to a car.

Any one of the foregoing, or better still a combination of them, then gives the lucky owner of these pieces of automotive trash *carte blanche* to build himself a Cobra, or get one of the specialists to do his work for him. Strangely the result of all this is very often for sale at some stage of the proceedings…

The excellent book *Shelby American World Registry* makes amusing and enlightening reading when one sees the shenanigans that various people have obviously got up to, some chassis numbers even having three versions of the car in existence. The supposed histories of some chassis numbers and the cars emanating from these numbers, not from Thames Ditton, make you marvel at the bare-faced cheek of it all.

It is patently obvious that certain people have imagined or had good reason to believe that a certain car no longer exists and have 'found' its remains. On occasions there have been red faces when the real thing puts in an appearance, but think of the other times when they have got away with it.

All this is not to deny that incomplete cars have surfaced in unlikely places and may continue to do so, or that there could be a perfectly good reason for this providing they have impeccable and irrefutable provenance. Any Cobra built up from such remains, given that the remains are of a reasonable substance, would of course have a legitimate claim to the chassis number pertaining to it.

The third and final post-Thames Ditton Cobra part of the story concerns the activities of a Mr Brian Angliss.

In the early 1970s this undoubtedly talented gentleman began to build the odd AC chassis and even a replica Cobra or two. Word of his prowess in this direction and the fabrication of pieces for Cobras soon got around and in time he founded a firm named Cobra Parts, based at Chessington in Surrey.

His early days in the Cobra business were recounted by Mike Taylor in *Sporting Cars* magazine for October 1982 and I quote: 'He began by restoring a 7-litre Cobra, took the measurements to help him build himself another car, and started to gain a reputation, with requests coming from America for spares. Cobra Parts were now in business. Anyone with a genuine log-book from an AC could have a car made. He shunned publicity and operated behind box numbers and ex-directory telephone

AFTER THE COBRA

One of the Autokraft creations pictured in 1983 on the old Brooklands track, close to the factory.

numbers, but people still sought him out. He still is not keen on publicity – we failed to persuade him to be photographed'.

Over the years Angliss befriended Derek Hurlock, senior surviving member of the family which had purchased the AC factory in 1930 when it was in the hands of the official receiver. Impressed by the quality of the work produced by Angliss and his staff, Hurlock was happy to allow him to carry on making copies of the various components which make up a Cobra. After all, AC no longer made the cars and Hurlock's interests were with the development of the ME 3000 and other work undertaken by the factory. For many years the Hurlock family had looked to activities other than the manufacture of motor cars for their main source of profitable revenue, these ranging from airport fire fighting vehicles to golf trolleys, and including, of course, the many invalid cars that they were contracted to produce by the National Health Service.

In due course Angliss was able to acquire from the AC factory some of the original body bucks used in the making of Cobra bodywork, and through this became the factory's official supplier of any panelwork or bodies for these cars. During 1978 he was approached by an American 'Cobraphile', Richard Buxbaum, who was prepared to come up with the money for him to construct virtually exact copies of the 427 to be sold in the US. The first of these, as they purported to be nothing other than new cars, failed to comply with American type approval regulations and consequently were rejected. The bold step was then taken to produce a modified car that would conform to the laws of that country, this apart from anything else involving the expense of crash testing.

At first the cars that resulted from all this could only go officially under the firm's name of Autokraft. For some reason best known to himself, Angliss called his car the 'MkIV' version of the Cobra. Why the 'MkIV' I would love to know as he should have been aware, as a result of his close association with AC, that there had never been a 'MkIII' as far as the factory was concerned. 'MkIII' is merely a retrospectively applied misnomer for the 427, first used by goodness knows who, that has been quoted so frequently that it has passed into folklore.

In the early 1980s Angliss had a breakthrough when he reached an agreement with AC Cars Ltd. Roger Bell reported this for the March 1982 edition of *Thoroughbred & Classic Cars* magazine and I quote: 'AC grant to Autokraft on an exclusive basis authority and licence during the period of this agreement to utilise and exploit the use of the trademark AC.

'In that snippet from an historic contract signed last month, you have the nub of the matter – the rebirth of one of the world's most famous and coveted classics, the AC Cobra, built by Autokraft bearing the AC badge. Legally speaking, it is not a Cobra because Ford of America now own that evocative name and it would need the consent of the US giant (which might not be forthcoming) to complete the famous nomenclature. But it's the use of the proper AC badge, available for the next 25 years, that sets the new Autokraft AC aside from all the other Cobra clones and replicas.'

At around the same time Angliss obtained permission from the Ford Motor Company to use the Cobra trademark, which it had bought from Shelby in 1967.

His next acquisition was more far-reaching. For some time Derek Hurlock had been thinking of finally retiring from the motor business. The original AC factory site had been sold for development and a move had been made to smaller premises. Then the rights to the ME 3000, of which the factory had produced fewer than 100 examples, had been passed in 1985 to a new and short-lived company in Scotland.

It was thus that Angliss was able now to become the

ESSENTIAL AC COBRA

owner, in the spring of 1986, of the company whose creations he had for so long cribbed.

The precise nature of who exactly owned AC Cars is a little complicated. Angliss and Autokraft became joint owners of AC Cars Ltd with William West's AC Holdings. West had purchased the Hurlock family's shares in the AC company and AC Holdings now had 51 per cent of the new company. But Angliss was definitely in charge.

Whatever the situation, Autokraft issued a leaflet at this time extolling its own virtues as manufacturer of the MkIV Cobra and also took pains to mention that it could undertake all types of projects from wheelchairs to the required number of cars in order for a manufacturer undertaking a competition programme to be able to meet homologation regulations. A grandiose new factory was planned, and would house both Autokraft and AC Cars under one roof with 25,000sq ft and 50,000sq ft of space allocated to them respectively.

During 1987 the Ford Motor Company bought AC Cars, or rather a 50.96% share of it, and the new factory began to be built on a site on the new Brooklands industrial estate. Work had begun on a new Ace, while the Autokraft side of the business still continued to produce MkIV Cobras. Angliss's liaison with Ford seemed to be an unhappy one and not a little vitriol was thrown, mainly, it seemed, due to problems regarding the projected new Ace. Whatever the real reasons for the discontent, Ford repeatedly applied to the English courts in 1990 and '91 to have the company wound-up but Angliss managed to fight this off. An out-of-court settlement was reached between the two parties during 1992 and Angliss became owner of the company.

MkIV Cobras were still available, even in lightweight form. The price structuring of the products, especially of the lightweights, at times appeared a little strange, and it might have appeared that they were sold for 'as much as you can get'.

In 1995 the company went into administrative receivership. The new Ace, with its multiplicity of styling and major structural changes during development, as well as the fact that in all probability it cost considerably more to make than it could earn, was certainly one of the reasons behind the financial problems.

During 1996 activities at AC Cars (*née* Autokraft) were temporarily suspended owing to receivership. Engineless and unpainted 'Glider' version of 427 (above) is supplied to the US in this incomplete state to avoid Construction & Use regulations, harking back to the days in 1965 when homologation 427s were shipped in similar form. This car carries chassis number CSX 4015, although it is a mystery what Carroll Shelby has to do with it considering what 'CSX' stood for then. Elsewhere in the factory a stack of chassis/body frames stands behind rear sections of aluminium bodywork.

As this book is being written the company lies in limbo with a skeleton staff finishing off a few orders, and half-completed Cobras, chassis frames and bodies keeping company with Mr Angliss's other passions – his collection of classic motorcycles and two half-restored Hawker Tempest aircraft. His palatial office upstairs is locked while two or three faithful administrative staff deal with any enquiries.

APPENDIX

Production figures

LEAF-SPRUNG CARS

Chassis numbers	Production
CSX 2000 to CSX 2074 (260 motor and steering box, including one RHD car – CS 2030 – made as a demonstrator for AC cars)	75
CSX 2075 to CSX 2125 (289 motor and steering box)	51
CSX 2126 to CSX 2589 (289 motor and rack and pinion steering, including two RHD cars – CSX 2130 and 2131 – and four further chassis – CSX 2286, 2287, 2299 and 2300 – built up as Daytona coupés. In addition, 23 Le Mans and FIA race cars were constructed within this chassis run)	464
CSX 2601 and CSX 2602 (289 motor, built up as Daytona coupés)	2
COX 6001 to COX 6062 (289 motor, European cars made in RHD [COB] and LHD [COX] form)	60
A 98 (289 factory coupé for 1964 Le Mans)	1
Total	**653**

COIL-SPRUNG CARS

CSX 3001 to CSX 3055 (all homologation cars, some converted by Shelby into 427 SC models; CSX 3001 left the AC factory in chassis form only)	55
CSX 3063 (96in wheelbase chassis)	1
CSX 3101 to CSX 3358 (427 road cars)	258
COB 6101 to COX 6127 (AC 289 cars, made in RHD [COB] and LHD [COX] form)	27
Total	**341**
Grand total (all Cobras)	**994**

Racing equipment

If it was the intention to use a leaf-sprung Cobra for racing, the Shelby organisation offered the following equipment:

Aviad sump (9-quart), oil cooler, 2 × 4v intake manifold, 3 × 2v intake manifold, 1 × 4v intake manifold, four Weber twin-choke carburettors and manifold, competition exhaust system, dual-coil ignition, Spalding 'Flamethrower' magneto, large-capacity radiator, aluminium radiator, aluminium crankshaft and water pump pulleys, aluminium engine block, aluminium cylinder heads, high compression pistons, Engle competition pushrods, differential oil cooler, bonnet scoop, front brake air scoops, rear brake air scoops, aluminium brake calipers, twin master cylinders, cold air box and intake, extra bonnet vents, undertray, 17-gallon fuel tank, 37-gallon fuel tank, electric petrol pump, large-diameter petrol filler, front anti-roll bar, rear anti-roll bar, competition springs, Koni adjustable shock absorbers, steering box bracket (worm and sector steering), rollover bar, competition wire wheels, alloy wheels and hubs, wheelarch extensions, close-ratio gearbox, competition seats, competition windscreen, radiator stoneguard, competition lights

Technical specifications

LEAF-SPRUNG COBRA, 260 & 289 (1962-64)

Engine V8 **Construction** Cast iron block and heads **Crankshaft** Five main bearings **Bore x stroke** 260 – 96.5mm × 73.0mm (3.80in × 2.87in); 289 – 101.7mm × 73.0mm (4.00in × 2.87in) **Capacity** 4261cc (260cu in) until chassis number CSX 2075, then 4738cc (289cu in) **Valves** Overhead valves operated by pushrods **Compression ratio** 9.2:1 (260), 11.6:1 (289) **Fuel system** Four-barrel Autolite or Holley carburettor **Maximum power** 260bhp at 5600rpm (260), 275bhp at 6000rpm (289) **Maximum torque** 269lb ft at 4500rpm (260), 314lb ft at 3400rpm (289) **Transmission** Four-speed Borg Warner T10, all-synchromesh **Gear ratios** 260, from *Road & Track*, September 1962 – 1st, 2.36 (8.36); 2nd, 1.78 (6.30); 3rd, 1.41 (4.99); 4th, 1.00 (3.54). 289, from *Road & Track*, June 1964 – 1st, 2.36 (8.90); 2nd, 1.78 (6.71); 3rd, 1.41 (5.32); 4th, 1.00 (3.77) **Final drive ratio** 3.54:1 on all cars with COB and COX chassis prefixes and on CSX chassis until CSX 2069, then 3.77:1 usually fitted **Top gear mph per 1000rpm** 21.8mph (3.54:1), 20.2mph (3.77:1) **Brakes** Girling discs front and rear **Front suspension** Independent with single transverse leaf spring, Armstrong telescopic shock absorbers **Rear suspension** Independent with single transverse leaf spring, Armstrong telescopic shock absorbers **Steering** Rack and pinion, but Bishop Cam steering box (as used on AC Aces) on early cars up to chassis number CSX 2125; 2 turns lock to lock, turning circle 34ft (10.4m) **Wheels/tyres** Dunlop 15in 72-spoke triple-laced wire wheels; 5½in rims until chassis number CSX 2159, and 6in thereafter; Goodyear 6.50/6.70 tyres **Length** 151.5in (3848mm) **Width** 61in (1549mm) **Height** 49in (1244mm) **Wheelbase** 90in (2286mm) **Front track** 51.5in (1308mm) **Rear track** 52.5in (1333mm) **Ground clearance** 260, 7in (178mm); 289, 5in (127mm) **Unladen weight** 260, 2020lb (916kg); 289, 2170lb (984kg) **Frontal area** 16.6sq ft (1.54sq m) **List price** $5995 (1964)

COIL-SPRUNG COBRA, 427 & 428 (1965-66)

As leaf-sprung Cobra except: **Bore x stroke** 427 – 107.4mm × 96.0mm (4.23in × 3.78in); 428 – 104.9mm × 101.1mm (4.13in × 3.98in) **Capacity** 6998cc (427cu in), 7014cc (428cu in) **Compression ratio** 10.4:1 (427), 10.5:1 (428) – but variations according to type of heads used **Fuel system** Two Holley or Autolite four-barrel carburettors **Maximum power** 480bhp (427), 350bhp (428) **Maximum torque** 460lb ft at 2800rpm **Transmission** Ford four-speed all-synchromesh **Gear ratios** From *Road & Track*, July 1974 – 1st, 2.32 (8.21); 2nd, 1.69 (5.98); 3rd, 1.29 (4.57); 4th, 1.00 (3.54) **Final drive ratio** 3.54:1 **Front suspension** Independent by unequal-length wishbones, coil springs, telescopic shock absorbers **Rear suspension** Independent by unequal-length wishbones, trailing arms, coil springs, telescopic shock absorbers **Steering** Rack and pinion, 2.8 turns lock to lock, turning circle 36ft (11.0m) **Wheels/tyres** Until chassis number CSX 3055 usually Hallibrand magnesium alloy, but GT40-pattern wheels (7½in front, 9in rear) on some later cars; from chassis number CSX 3101 'Sunburst' 10-spoke wheels but some cars between CSX 3125 and CSX 3157 again had GT40-pattern wheels (both types 7½in front and rear); all wheels 15in diameter and fitted with Goodyear tyres **Length** 156in (3962mm) **Width** 68in (1727mm) **Track** 56in (1422mm) **Ground clearance** 4.4in (112mm) **Unladen weight** 2529lb (1147kg) **List price** $7495 (1966)

COIL-SPRUNG COBRA, 289 (1966-68)

As 427 & 428 except: **Bore x stroke** 101.7mm × 73mm (4.00in × 2.87in) **Capacity** 4738cc (289cu in) **Compression ratio** 11.0:1 **Fuel system** Autolite four-barrel carburettor **Maximum power** 271bhp at 6000rpm **Maximum torque** 312lb ft at 3400rpm **Transmission** Four-speed Borg Warner T10, all-synchromesh **Final drive ratio** 3.54:1, 3.31:1 after chassis number COB 6120 **Wheels/tyres** 6-15in Dunlop wire wheels **Unladen weight** 2398lb (1088kg) **List price** £2952 (1967)

ACKNOWLEDGEMENTS

Special colour photography in this book includes cars owned by Robin Stainer (260), Herb Wetanson (289), Nick Green (289) and Anthony Posner (427) – this work was variously carried out by Rinsey Mills, Jim Feldman and Mick Walsh. Other sources of photographs were *Road & Track* magazine (thanks to Otis Meyer), Geoff Goddard (thanks to Doug Nye), Neill Bruce (including the Peter Roberts Collection), Quadrant Picture Library, the National Motor Museum, David Hodges, LAT Photographic, Ludvigsen Library (pictures by Stanley Rosenthall), Haymarket Publishing Motoring Archive, Phipps Photographic, Ford, Ted Walker (Ferret Fotographics), EMAP National Publications Limited, Ned Scudder, Laurie Caddell, Eberhard Kittler, Trevor Legate and Rinsey Mills.

Postscript

In December 1996 AC Cars, Britain's oldest surviving car manufacturer, was saved from extinction. The benefactor is Alan Lubinsky, President of Pride Automotive Group Inc. A new subsidiary company, AC Car Group Limited, has acquired the assets and business of AC Cars. This is exciting news for all lovers of the marque.